ADDING BIOLOGY

In Soil and Hydroponic Systems

ADDING BIOLOGY

In Soil and Hydroponic Systems

For Conventional, Sustainable and Organic
Plant Growing Systems

*Outdoor Applications and Indoor Controlled
Environments
In Soil and Soilless Media*

***Elaine R. Ingham**, Ph.D.*

***Carole Ann Rollins**, Ph.D.*

Sustainable Studies Institute, Corvallis, Oregon
Nature Technologies International, LLC, Novato, California

For information:

Sustainable Studies Institute
1750 SW 3rd Street, Suite K, Corvallis, OR 97333-1796
Phone: 541-257-2614 Fax: 541-752-5142
www.sustainablestudies.org
Email: info@sustainablestudies.org

Nature Technologies International, LLC
14 Tilden Drive, Novato, CA 94947
Phone: 415-898-5895 Fax: 707-940-0444
www.nature-technologies.com
Email: naturetech@earthlink.net

Printed in the United States of America

First Printing May 2006

Revised May 2008

ISBN 0-9797561-0-3 (paperback)

TABLE OF CONTENTS

Preface

Commercial agriculture and non-agriculture consumers are sensitive to environmental impact of toxic chemical pesticides and synthetic fertilizers on soil, soilless media and plants. With government regulation and gradual removal of toxic pesticides currently and in the future, alternative non-toxic substitutes are desperately needed. The same chemical companies losing revenue from toxic pesticides sales are gearing up biotech production of disease resistant seeds utilizing genetically modified organisms (GMOs) to refill that market niche they once held. But unknown consequences of these potentially harmful GMOs and introduction into the marketplace before adequate testing, has caused controversy and concern among potential consumers. This situation is providing a unique opportunity for non-toxic alternatives to be introduced into the marketplace and fill these gaps.

Current research from Dr. Graham Lancaster from Southern Cross University in Australia and Dr. Elaine Ingham, a microbiologist from the Soil Foodweb Labs reported in 2004 that every soil examined contains enormous amounts of total nutrients, but they are not available to plants in their current form. Beneficial micro-organisms must be present in the right ratios for the specified plants to make these nutrients available to the plants at rates the plants require. If organisms are not present they must be added through inoculums such as high quality compost tea. Plants provide foods to keep the organisms performing their functions. The organisms must also be fed properly through the times that plants are dormant by adding compost or other types of nutrients.

There are many types of biological products for soil and soilless media enhancement available on the commercial market today and more entering daily. New research, new formulations, new stabilization techniques are enabling the expansion of these product to fill the increasing demand from consumers wanting alternatives to toxic chemical pesticides and synthetic fertilizers.

This notebook is an attempt to provide basic information about adding biology to soil and soilless media whether in outdoor fields or indoor controlled environment hydroponics systems. Once we are equipped with the knowledge, we can then make intelligent decisions when faced with so many choices of brands and products in the marketplace. Whether we are using synthetic fertilizers/nutrients or sustainable practices, or have converted to organic systems, there is a way to add biology to enhance production, yield and quality. This notebook will provide you with some of the parameters, tools and knowledge so you can integrate biology into your specific growing system.

This notebook was compiled from information introduced at a Hydroponics and Compost Tea Seminar given by Dr. Elaine Ingham at the Sustainable Studies Institute in Corvallis, Oregon on December 3, 2005. Since many of the concepts were also applicable to soil-based systems, the reference to soil systems were added to the text, when appropriate. This notebook is not a complete investigation and reporting of the different types of biology that can be added to hydroponics or soil systems. The authors acknowledge there are different biological systems in use around the world with varying types of biology recommended, that are not addressed in this notebook. Such an investigation would be an extensive research project beyond the scope of this undertaking.

This is an attempt to provide a user-friendly knowledge base in non-technical terms, serving as a basic introduction only, to a broad scientific field of study. Science is only beginning to understand this last frontier of the scientific universe, the biological microscopic communities providing nutrient cycling, plant protection and plant growth mechanisms. If you want more in-depth knowledge, Elaine Ingham, PhD provides several types of services, from private consultations to educational workshops and microscope courses through the Institute of Sustainable Studies in Corvallis, Oregon. CDs and books by Dr. Ingham are also available through Nature Technologies International, LLC and Earth Fortifications. Microbiological testing services are available through Soil Foodweb Laboratories throughout the world. A listing of laboratories and products available along with addresses are provided in the back of this book for those interested in further information. Carole Ann Rollins, Ph.D. provides several types of services, from private consultations to group seminars, lectures, and licensing of regional compost tea brewing centers through Nature Technologies International, LLC in Novato, California. Books by Dr. Rollins are available through Nature Technologies International, LLC and Earth Fortifications.

Acknowledgements

Thank you to Kevin Haines and Gladys Rollins for proofreading and editing. Thanks to Steve Diver for suggestions. Thanks to Tina Peterson for gathering photos from their projects and providing descriptions. Thanks to Tom Alexander for the encouragement to write the text and the title. Thanks to Joe Whaley and Matt Slaughter for continued support. Carole thanks James Eddington for all the unending and unwavering encouragement and support through the years.

Elaine R. Ingham, Ph.D. Corvallis, Oregon
Carole Ann Rollins, Ph.D. Novato, California

1. INTRODUCTION

There are fundamentally two types of systems for growing plants in soil and soilless media:

1. Inorganic/Synthetic and
2. Organic[1]/Biological

The basic difference between the two systems is that an organic system requires carbon[2] (meaning sugars, proteins, and other cellular building blocks), processed by organisms, while inorganic systems attempt to provide the nutrients in inorganic forms that provide all the organic materials it will need, and since no one single plant species can provide all these products, inorganic systems flounder.

Both systems have the goal of growing plants with as little effort as possible. Inorganic systems can give financial returns for the short term, but there are long term consequences that are ignored with the inorganic approach. In the long term, the organic approach is the only way to sustain production. The full benefits and costs of organic versus inorganic systems needs to be examined in relation to the impacts on water quality, soil quality, diseases, pests and food quality.

INORGANIC AND BIOLOGICAL APPROACHES: *We can attempt to provide each nutrient at the right concentration at the right time in the growth cycle of the plant (inorganic approach) or use biology, which natural systems have been developing for eons of time to cycle nutrients into the right forms, at the right times, as controlled by the plant (biological approach). The photos above show some of the different types of growing environments that choose to use the inorganic or biological approach. (Photos courtesy of Nature Technologies International, LLC and Sonoma Mountain Institute).*

There is a significant difference in approach between organic and inorganic, but both have the goal of growing plants rapidly, inexpensively, and with maximum benefit to the grower and consumer.

In order to grow healthy plants we need to be aware of the nutrients and minerals plants need, and adequately meet those needs within the growing system used. In both inorganic and organic systems there is a need to provide each nutrient in the right concentration at the right time in the growth cycle of the plant. With inorganic systems, it becomes necessary to artificially manage what biological natural systems have been performing successfully for eons of time. Plants control the biological approach and work with the biology to deliver the right nutrients to the plant in the right forms at the right times.

Types of Nutrients for Growing Plants

1. Non-Mineral Nutrients -- Hydrogen (H), Oxygen (O), Carbon (C)
2. Primary Mineral Nutrients or Essential Macronutrients -- Nitrogen (N), Phosphorus (P), Potassium (K)
3. Secondary Mineral Nutrients – Calcium (Ca), Magnesium (Mg), Sulfur (S)
4. Essential Micronutrients -- Boron (B), Copper (Cu), Iron (Fe), Chloride (Cl), Manganese (Mn), Molybdenum (Mo), Sodium (Na), Nickel (Ni), Zinc (Zn)
5. Beneficial Minerals -- Cobalt (Co), Silicon (Si)
6. Trace minerals -- Over 100 types

Plants Take Up Nutrients in the Form of Ions

Regardless of whether a system is biological or synthetic, plants need nutrients in the form of ions[3]. An ion is a positively or negatively charged compound.

Salts, for example, are made up of ions and are soluble if they can easily be dissolved in water[4]. That means when the salt is added to water, the hydrogen (H^+; positive charge) portion of water will complex with the negative portion of the salt and the hydroxide (OH^-; negative charge) portion of the water molecule will complex the positive part of the salt.

Examples of nutrient salts include ammonium chloride, where an ammonium ion (NH_4^+) is the positive part of the molecule, and a chloride ion (Cl^-) is the negative part of the salt. In calcium carbonate, the calcium ion has twice the positive charge (Ca^{2+}) and so requires two carbonate ions (CO_3^-) to form the molecule. Iron can complex with negative ions to form salts, although not as a soluble. Generally, iron (Fe^{2+} and Fe^{3+}), magnesium (Mg^{2+}), potassium (K^+), and sodium (Na^+) are all clearly capable of forming salts with the right kinds of negative compounds (called anions), such as chloride or hydroxide.

Salts can also form when different materials mix together and combine to form compounds that have the positive-negative parts of the compound. If water can "pull them apart," then the salt is a soluble salt. If water cannot "pull them apart," the salt is not soluble.

Why do salts do this? Because of the simple positive and negative charge interactions. All salts have been put in this category because they behave this way. For example, the inorganic fertilizer ammonium nitrate is split apart during the process of being dissolved in water. Ammonium (NH_4+) interacts with hydroxide and Nitrate (NO_3-) interacts with hydrogen. The inorganic fertilizer thus dissolves.

But think about what that means for water availability in the soil or soilless media. For each molecule of inorganic nutrient or fertilizer added, one molecule of water has been "taken away" from the plant, or from the microbes. Negative impacts on microbes can be severe if the concentration of inorganic nutrients or fertilizers are 100 pounds per acre or more. Plants aren't typically harmed until the concentration of inorganic fertilizers or nutrients reaches a much higher level. Plants are tolerant of desiccation (lack of water) to a much greater extent than many of the beneficial bacteria and fungi.

Once dissolved, the ions (positive or negative charged ions complexed with water) can then be taken up through the process of diffusion into the plant cells. Most plants take up their major nutrients through simple diffusion of ionic forms.

If the plant takes nutrients up by simple diffusion, how can nutrients become more concentrated in the plant than in the soil or soilless media? The plant converts those nutrients into a different form once the nutrient is inside the cell. For example, the plant takes nitrate (NO_3), re-arranges it into an amide (NH_2) and complexes it onto a straight chain carbon backbone (C-C-C for example) to make a protein (C-C-C – NH_2). Because N is no longer present as nitrate, however, simple diffusion will continue to bring more nitrate into the root.

Chelation of Nutrients

Salts (made up of ions) can be further processed so they can diffuse more easily into plant cells. A process called chelation[5] can make diffusion easier. Chelation can be induced chemically or biologically. Chelation means that each ion is bonded to a protein (made up of amino acids). Chelation can happen biologically as part of an organism's digestive process or even a plant's metabolic process. Chelation can also be induced chemically during the manufacturing process. Chelated nutrients are no longer salts. When a chelated nutrient is available, it takes much less energy for the plant to take up, because it is absorbed more efficiently and simply.

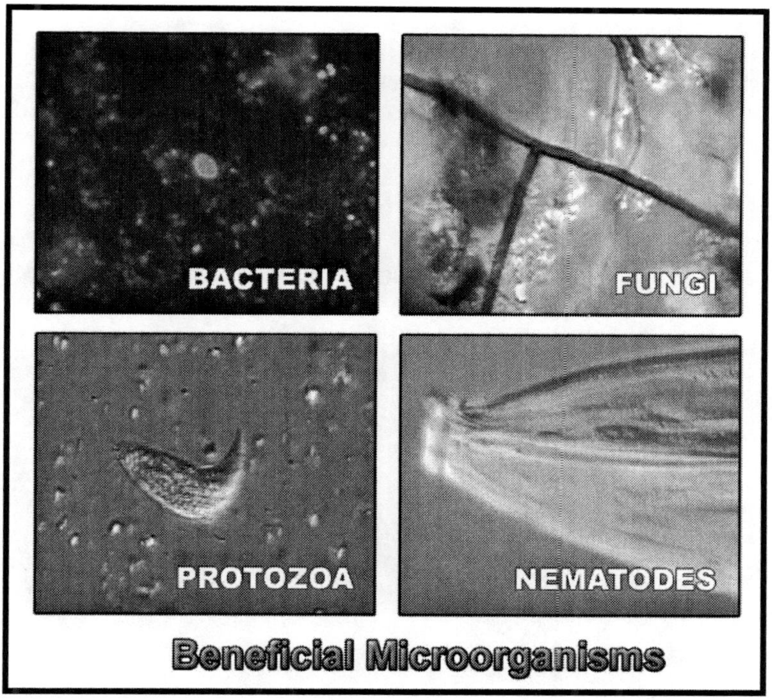

BENEFICIAL MICROBES: *Four categories of beneficial microorganisms, i.e., bacteria, fungi, protozoa and nematodes, are required for nutrient cycling to maintain a balanced organic/biological soilless growing environment. Conditions in the hydroponics solution, the growing medium, compost, or soil must select and promote the growth of beneficial organisms, not the disease organisms. Alter conditions in the medium and make sure the beneficials, and food, are present to bring about those conditions.* (*Photo courtesy of Soil Foodweb, Inc.*)

Beneficial and Non-Beneficial Microorganisms

Billions of microbes naturally occur everywhere on our planet – on our plants, our food, in our soil and soilless media, in water, on our clothes, even on and in our bodies. In the plant growing world, some of these are good microbes because they produce materials or create conditions that are beneficial for plant growth and health. Beneficial microbes also interact with other microbes in a way to enhance nutrient cycling. Some microbes are non-beneficial because they can cause disease and make waste materials (metabolites or phytotoxic compounds) that are harmful to plants and other beneficial microbes. While some of these harmful waste materials may, in fact, kill disease-causing organisms, they can also harm plants and other beneficial organisms in the process.

The Soil and Hydro Food Web

In soil and soilless growing systems there are different types of biological organisms or microbes that are invisible to the eye, and hence are called microorganisms – bacteria, fungi, protozoa and nematodes. All of these organisms can live in soil and soilless media, including water.

Microarthropods are another type of organism that can also be found in soil and some types of soilless media, such as coir, peat, or rock wool. These organisms hold onto nutrients and cycle them into a plant-available form, directly or indirectly through interaction with other organisms.

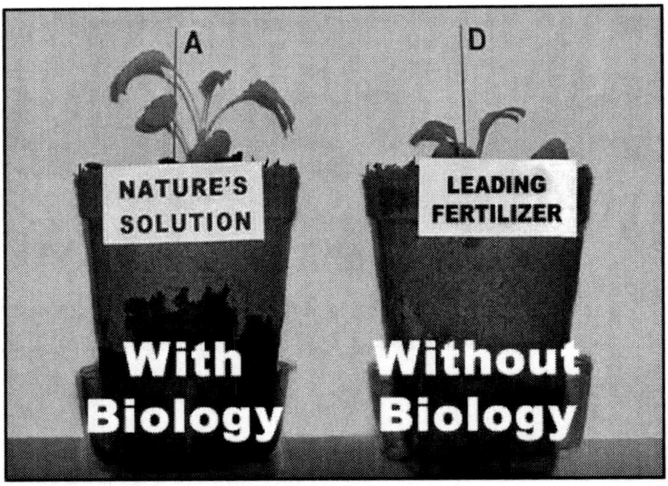

RADISH TRIAL: *This was a controlled study of radish seeds grown with biology in the form of compost tea used in the left pot and inorganic synthetic nutrients used in the right pot. Clearly the plants in the system with beneficial biology grew more rapidly and were healthier than the system without the necessary beneficial organisms (Photo courtesy of Nature Technologies International, LLC.).*

Energy and nutrients are converted throughout this nutrient cycling process as one organism eats a nutrient and retains it, then another organism eats the first organism. The nutrients eaten by the original organism then get released to feed the plants. The waste material created as a byproduct of the microorganism's metabolism is processed and excreted into an undigested material that is immediately available for plant uptake as nutrients needed for growth. Sometimes these waste materials are in the form of enzymes or bound with protein into a chelated form. These microorganisms also perform other functions, such as aiding the decomposition of plant matter and breaking down toxic materials produced by plants, plant residues, heavy metals and pesticides.

SYNTHETIC SYSTEMS: *Synthetic systems utilize these inorganic compounds that are in a plant-available (inorganic) form. (Photo courtesy of Nature Technologies International, LLC.)*

2. SYNTHETIC OR INORGANIC SYSTEMS

In synthetic or inorganic soil or soilless growing systems, nutrients are strictly in an inorganic salt-based ionic form, such as ammonium, nitrate, potassium nitrate, gypsum (calcium sulfate), etc. The nutrients are mixed together in precise amounts and added to these growing systems. The nutrients are comprised of inorganic compounds (macronutrients, micronutrients, and trace minerals) that have been extracted and purified from a source material, and are usually chemically processed. Synthetic systems utilize these inorganic compounds that are in a plant-available (inorganic) form. Inorganic compounds dissolve in water or are attached to minerals. These soluble nutrients can enter the plant cell wall through simple diffusion. By bonding these nutrients with a protein (the process of chelation) they can more efficiently be utilized and taken up by the plant. It takes much higher concentrations of nutrients if there is no chemistry or biology to help the nutrients become chelated.

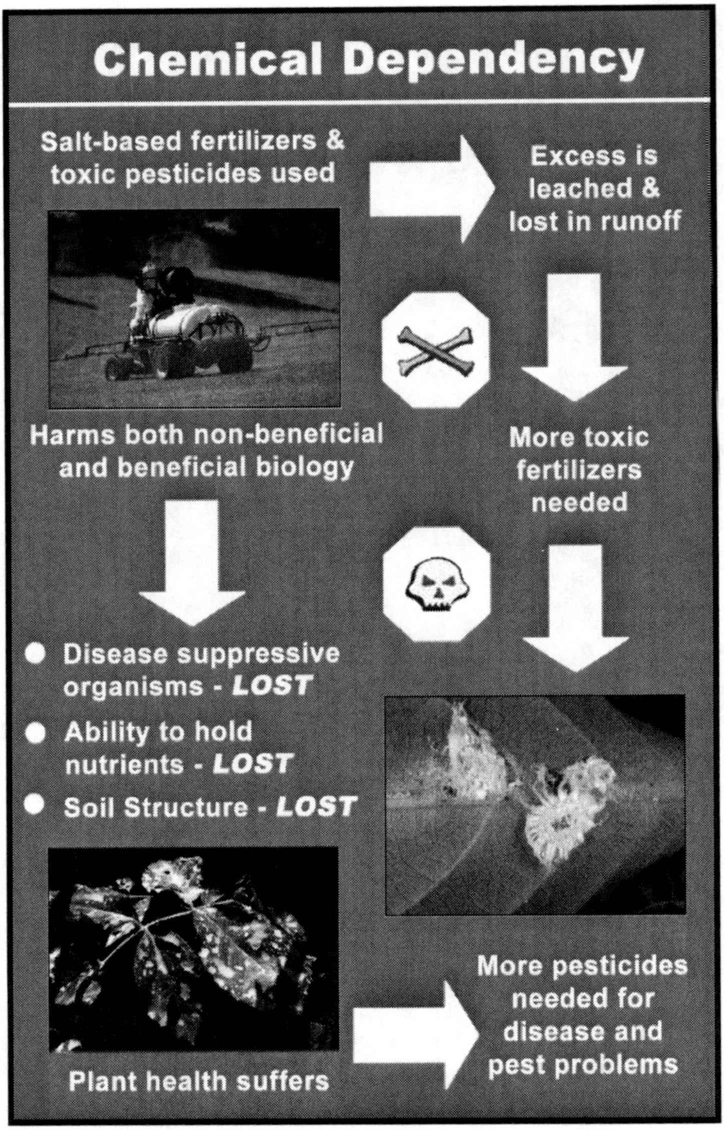

CHEMICAL DEPENDENCY: *When we use salt-based fertilizers and toxic pesticides, these harm both the non-beneficial and beneficial biology. Organisms are responsible for a great deal of the nutrient-holding capacity of the soil. The organisms in soil can have a greater nutrient-holding capacity than clay (consider,* 10^{12} *bacteria and the ability of 300 miles of fungal surfaces to hold nutrients in healthy root soil). When salt-based fertilizers are used that kill organisms, consider the loss of ability to hold nutrients. The excess nutrients are also leached and lost in runoff. Then more toxic fertilizers are needed to replace those lost and plant health ultimately suffers. The cycle continues to try to improve plant health and more pesticides are needed to combat disease and pest problems. (Diagram and photos courtesy of Nature Technologies International, LLC.)*

NO MORE PESTICIDES OR SYNTHETIC FERTILIZERS

BIOLOGICAL SYSTEMS: *Brian George of B. George Nursery in Novato, California has been achieving good biology by using compost tea, compost and worm castings on his two acre nursery for over three years. Brian uses biological products on virtually every new plant that comes in. Brian says, "I hate losing plants and really like them to look good. No chemical fertilizers or pesticides, only actively aerated compost teas have been used. While there are small amounts of rust and black spot, the 2500 roses I have in my inventory are the best I have ever seen, and the aphids and whiteflies have disappeared." (Photo courtesy of Nature Technologies International, LLC.)*

Chemical Dependency

Beneficial organisms in soil or soilless media are very important for nutrient retention and prevention of leaching. They retain a significant amount of nutrients on their surface, such as 10^{12} per gram, and 30 to 100 miles of fungal hyphae in a gram of root zone medium. If these organisms are killed by the use of toxic pesticides or high levels of inorganic fertilizers, they are unable to retain nutrients, build soil structure, or perform disease suppression functions. There is significant documentation through scientific study that the correct set of organisms in soil can be highly disease suppressive. The challenge lies in ensuring that the beneficials are selected and maintained.

BENEFICIAL AEROBIC MICROORGANISMS: *Beneficial aerobic microorganisms require food, air, and water to survive in soil and hydroponic systems. Dissolved oxygen levels must be maintained above 6 ppm (6 mg of oxygen per Liter) for the aerobic organisms to live. (Photo courtesy of Nature Technologies International, LLC.)*

3. BIOLOGICAL/ORGANIC SYSTEMS

Biological/organic nutrient cycling systems have been in existence ever since predators of bacteria and fungi came into existence about three billion years ago (to the best of our knowledge). Biological processing of nutrients into the proper "plant-available" form was not understood until a few years ago. With this new understanding we learned that biological cycling systems combine the best of chemistry, physics and biology with sound soil and hydroponics management practices.

Biological systems address and solve disease and insect pest problems through inhibition, competition and consumption of non-beneficial organisms, rather than merely masking the symptoms with poisonous chemicals. Poisonous toxic chemical solutions do kill the diseases or pests, but also kill the beneficial biology, which in turn hinders the natural nutrient cycling system from doing its job. It is not a question of whether this organic/biological system can work, but rather a question of HOW to make it work under various environmental conditions for different plant types and growing systems. Research in this area is accelerating rapidly.

Biology is required to transform nutrients into plant-available forms

<u>How the Biology Works</u>

Step One: Determine if significant amounts of any biological group is missing through reliable soil testing. The missing groups, or parts of groups, have to be added.

Step Two: Is any plant-available nutrient pool missing or low? If biology has been lacking for some time, soluble pools will be low and exchangeable pools certainly will be incorrectly balanced. The addition of inorganic materials can severely modify the plant growing habitat rapidly, killing biology. Whether liquid or solid, the missing nutrients need to be replaced. If last minute improvements are required a final application of chemicals (preferably chelated) can be added before restoring biology. With careful planning the biology can be balanced and needed pools brought back up to strength.

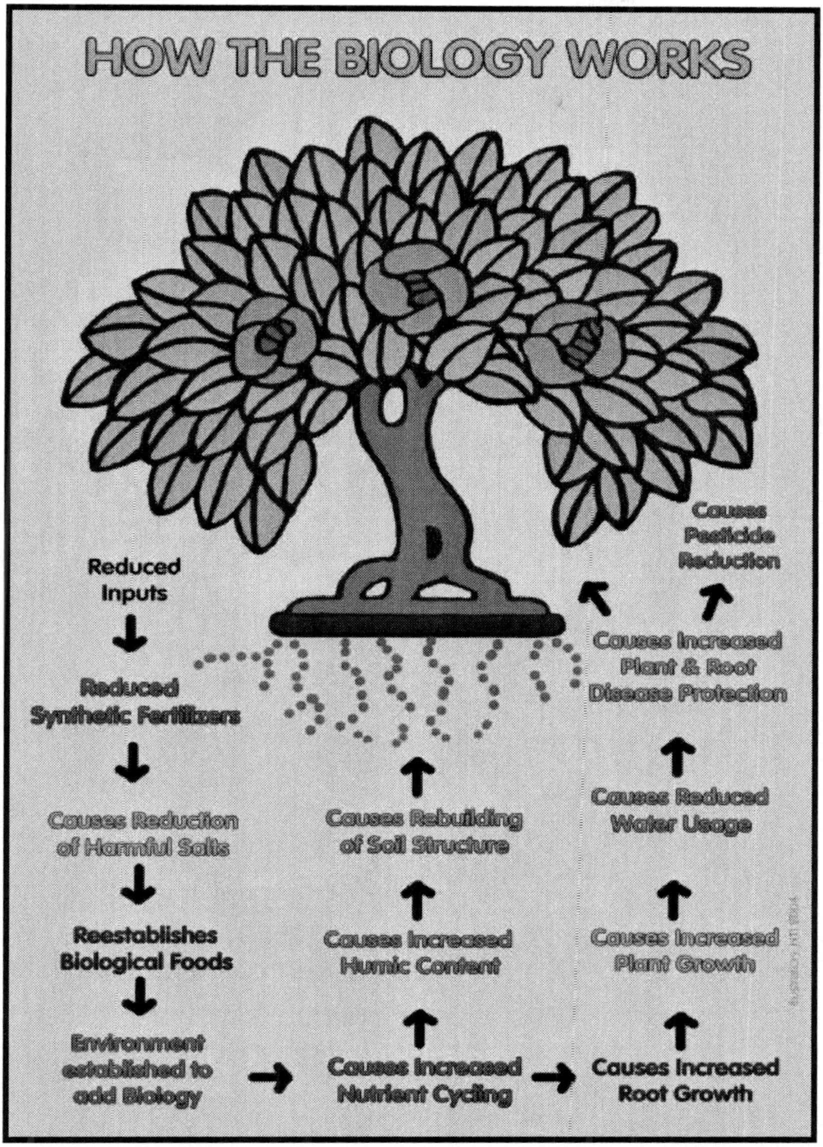

HOW THE BIOLOGY WORKS: *Step one is to determine if significant amounts of any biological group is missing. Second, determine if any plant-available nutrient pools are missing or low. The reduction of inorganic or chemical based inputs is necessary which includes synthetic fertilizers that contain harmful salts. Then biological foods must be introduced so that when biology is added it can be maintained. Once the biology has been added, this then causes the nutrient cycling to be increased, which causes the humic content of the soil to increase also. All of this enhances the rebuilding of the soil structure and causes plant and root growth and protection to increase and water usage and pesticides to be decreased. (Diagram courtesy of Nature Technologies International, LLC.)*

PLANTS FEED MICROBES

Exudates of Protein, Carbohydrates and Sugar

"Cake and Cookies"

CAKE AND COOKIES: Roots release "Cakes and Cookies" to feed microbes, to protect the root system from diseases and pests, to enhance nutrient cycling, and build structure to allow roots to grow through soil with good air passageways to help them. *(Diagram Courtesy of Soil Foodweb, Inc.)*

Organic Systems Require Nutrient Cycling

Organic systems require carbon in the form of sugars, starches, amino acids, proteins, lipids, etc. Carbon in these diverse forms can be produced only by organisms that make organic compounds. Organic compounds include both the living organisms, as well as their waste products and dead bodies. Macronutrients, micronutrients, and trace minerals are included in the structure of organic compounds in the proper ratios to make living plant and animal bodies.

Kelp, humic acid, bat guano, hydrolyzed fish, compost, worm castings and oyster shells, etc. are specific types of organic nutrients. These organic nutrients are not in ionic or salt forms, so they cannot dissolve in water to enter a plant through simple diffusion. Biology is required to transform these organic nutrients into plant-available forms.

Soluble inorganic nutrients do not require this biological nutrient cycling, but immobilized inorganic nutrients do require biological transformation in order to process them into the soluble forms. These nutrients in the "non plant-available" forms, for example, are tied up in rock phosphate, rock dusts, bedrock, sand, silts and clays, or are in non-soluble salt forms, such as hydrides, carbonates, or sulfates.

HEMLOCK TREE: *"This is a hemlock tree which had a very severe adelgid infestation (similar to an aphid) and now is covered with cones -- something not seen in years. This tree received several compost tea applications, both foliar and drench over the past year. The hemlocks received the same treatment as other trees with some additional drenching. We add 30 gallons of tea to 200 gallons of water with some fish, seaweed and humates and inject into the root zone at about 30-40 gallons per tree." Alan Magan, Chesapeake Gardens, Maryland. (Photo Courtesy of Alan Magan).*

Nutrient Cycling Depends on Biology

Beneficial microorganisms are an essential component of this nutrient cycling in a soil or soilless growing medium. Microorganisms function, and are needed, in pure water culture hydroponic systems, as well as soilless media like coir and rock wool. Bacteria and fungi (two types of microorganisms) take up nutrients that plants exude (release) for the express purpose of feeding microbes around their root systems or decomposing plant material. Microbes can use ionic or organic forms of nutrients.

Chelated forms of nutrients are immediately available for plant uptake. Nutrient additions can be reduced in a biological system because nutrients are not lost as a result of leaching, and the nutrient cycling process is more efficient than in inorganic, salt-based systems.

Plants also provide microbes with the source of energy (food in the form of sugars, proteins and carbohydrates) to perform chelation. These energy/foods are provided in the plant roots, on leaves, and all plant surfaces. Stressed plants do not provide enough energy/food for the microbes to proliferate like healthy plants do.

Chelated forms of nutrients are immediately available for plant uptake.

Reduction of Salt and Toxic Levels Essential

The salt concentration of inorganic nutrients is usually so high that addition of these forms of nutrients into a biological system usually kills many of the beneficial microorganisms by the dehydrating action of the salts. Salt concentrations above 0.1 to a maximum of 0.5 millisiemen (mS) (approx. 100-350 ppm) take water away from the organisms, as well as plants. This hinders the natural cycling of nutrients. However, when inorganic nutrients are chelated, salts have been altered to no longer cause these types of problems. Chemical based pesticides and herbicides, chlorine and chloramines, heavy metals, nitrate, sulfur, tannic acid, and carbonates in water are also harmful to microorganism populations. Heavy metals are toxic to organisms at levels of 3 to 10 ppm (0.01 to 0.02 mS), killing critical portions of the biological groups in soil and soilless media.

HEALTHY SOIL AND UNHEALTHY SOIL: *True soil or growing medium containing the correct set of organisms for the desired plant, will improve root health, root depth, water holding capacity and aerobic conditions. The photo on the top shows healthy soil and the two photos on the bottom show unhealthy soil conditions without biology. (Photo courtesy of Nature Technologies International, LLC.)*

Benefits of Biological Organic Systems

The right biology enhances disease protection, practically ending the need for pesticides. Bacteria and fungi retain the greatest amount of nutrients in their biomass, thereby ending the leaching of nutrients from soil and soilless media. As bacteria and fungi are consumed by protozoa, nematodes and microarthropods, biologically transformed nutrients are freed in the right forms, in the right places, and at the right times to feed the plants.

Bacteria and fungi decompose toxins and plant residues. As a result, thatch and un-colonized plant tissues are no longer present and no longer offer sites for pathogens and pests to over winter and survive.

Bacteria, fungi, protozoa, nematodes, microarthropods and earthworms help maintain structure in planting media, thus improving root health, root depth, water holding capacity and aerobic conditions. Bacterial glues, fungal threads, and arrangement of aggregates by protozoa, nematodes and microarthropods are all mechanisms which contribute to improving, building and maintaining soil and soilless media structure.

In inorganic/synthetic systems nitrogen, in the form of nitrate, encourages disease-causing organisms. In organic/biological systems beneficial microorganisms convert nitrogen from nitrate to biomass. High concentrations of nitrate encourage harmful disease-causing organism growth, while conversions of this form of nitrogen into protein or other biological forms means the disease-causing organisms are not encouraged.

Beneficial organisms occupy the space and use the food that would otherwise be present for, and enhance pathogen growth. Yield, quality, food nutrition, and profit potentials in organic/biological plant growing systems are thus maximized.

Yield, quality, food nutrition, and profit potentials in organic/biological plant growing systems are maximized.

4. PLANT GROWING SYSTEMS OUT OF BALANCE

In our current toxic, chemical dependent plant growing systems, degradation of environmental conditions have altered populations of microbes. As a result, things have become out of balance. When high concentrations of soluble nutrients are added to inorganic systems (because nutrients are not held in these soil and soilless media due to the loss of organisms which normally hold them) the stage has been set for disease-causing organisms to win the competition with the beneficials.

With the use of fungicides, nematicides, insecticides, herbicides and other toxic chemicals, competitors to diseases have been killed, and do not come back as rapidly as the disease organisms. Disease-causing organisms have a "boom-and-bust" life history strategy of producing high numbers of offspring in a short time, while the beneficial organisms grow more slowly. If even one beneficial organism survives a toxic chemical application, they may continue to thrive for a long time, but do not reproduce often, thus maintaining low populations. In this way, the use of toxic chemicals selects for disease-causing organisms instead of beneficials. Organisms that perform nutrient cycling (the process of making nutrients available to plants) are lost with the use of high levels of inorganic chemicals which in turn forces dependence upon inorganic nutrient inputs and pesticides.

Since high levels of toxic chemicals are required in an inorganic growing system, the resulting soil, wastewaters, soilless media, and plant materials cause pollution from these systems when discarded. Pesticides, chloramine and chlorinated water also all harm, destroy and hinder beneficial microbial growth, leaving conditions ripe for harmful pathogen (disease-causing) microbe populations to thrive. When this occurs we start to see problems with our plants – their growth, appearance, condition, health and yields. If diseases and pests abound, healthy plants are not possible.

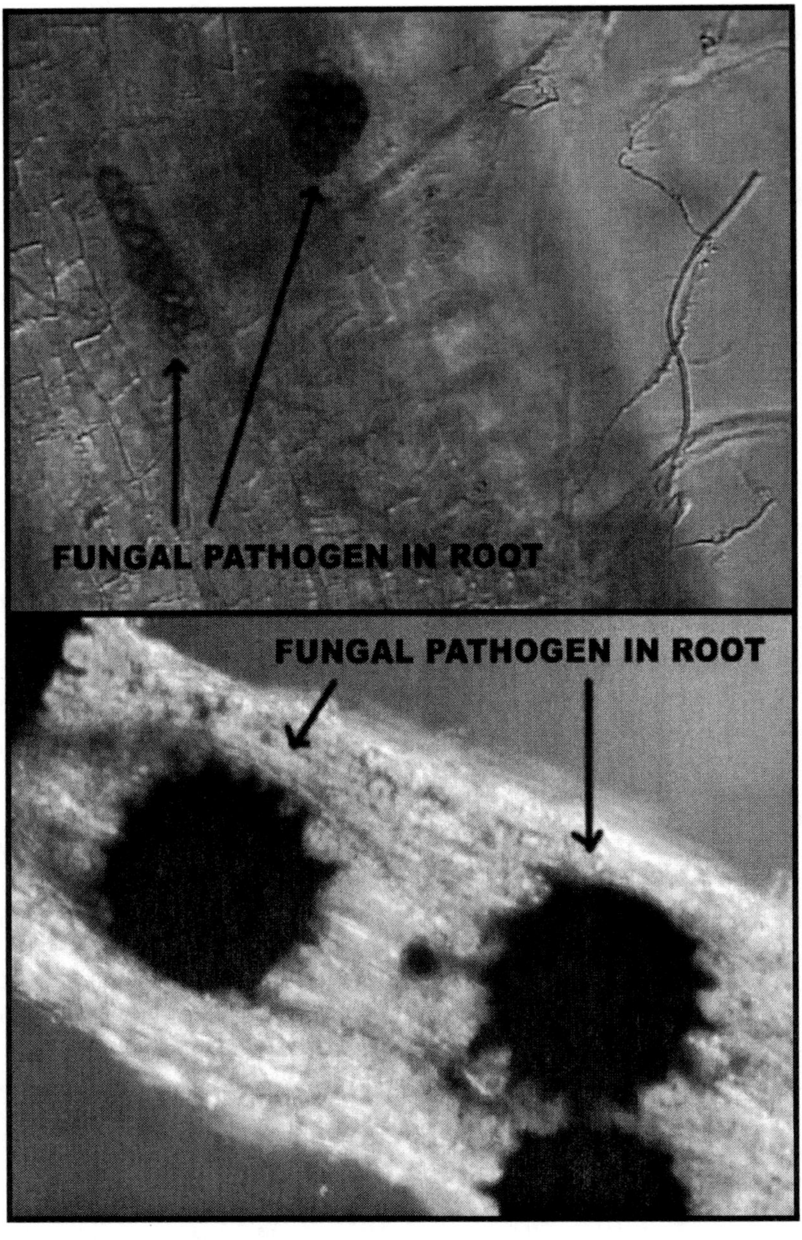

PATHOGEN PHOTOS: *Non-beneficial microbes, such as disease-causing pathogens, shown in the photos above as they occur in a plant root, are easy to distinguish in these spore stages. These organisms grow in conditions that are not beneficial for plant growth and health. (Photo courtesy of Soil Foodweb Inc.)*

Chemical Answers

Solutions to diseases, pests and lack of nutrients in soils and soilless media have been chemical based over the last 50 years, but it has been a "Band-Aid" approach. It may get rid of the immediate problem, but leave in its wake a more degraded environmental condition. This "Plant Chemotherapy" approach suppresses the immune system and thereby creates a whole new set of problems. The chemical approach kills pathogens or insects, but ultimately destroys beneficial organisms.

Toxic chemicals select for disease-causing organisms

This "Plant Chemotherapy" approach dramatically hinders the natural nutrient cycling system, so we become dependent upon more chemical inputs in order to grow crops and have good looking plants in our landscapes and growing systems. We have been offered few alternatives, other than chemical synthetic inputs from major manufacturers, consultants and advisors now for many years.

The chemical approach kills pathogens or insects, but ultimately destroys beneficial organisms.

50 Years Ago

50 years ago when we first started using toxic chemicals, the growing medium contained the beneficial biological life to grow plants at that stage of succession. But as tillage was used, bacterial growth was enhanced and organic matter was consumed rapidly by the bacterial bloom. As food diminishes, the disease-competitive and nutrient retaining beneficial organisms no longer can perform their function, so pesticides and inorganic fertilizers are needed. 50 years ago there were whole communities of beneficials in adjacent properties that would simply migrate back into impacted areas so we never noticed anything out of balance. Today, there are no communities next door, so they cannot migrate over. The only way to get things back in balance is to introduce the full spectrum of beneficial biology back into our growing systems through inocula such as occur in properly made compost, worm castings and compost teas.

50 YEARS AGO: *50 years ago when we first started using toxic chemicals, the growing medium contained the beneficial life to grow plants at that stage of succession. (Photo courtesy of Nature Technologies International, LLC.)*

This chemical dependency cycle can be ended. The natural nutrient cycling system can be re-established.

Biological Alternatives

This chemical dependency cycle can be ended. The natural nutrient cycling system can be re-established. Conditions can be changed so beneficial organisms proliferate, not disease pathogens, pests or problems. Beneficial microbes, instead of chemicals, can be used to "jump-start" nutrient retention, nutrient cycling, residue decomposition, soil and soilless media structure building and maintenance. These processes and organisms need to be monitored to make sure that the natural nutrient cycling processes develop and are maintained.

Conditions can be changed so beneficial organisms proliferate, not disease pathogens, pests or problems.

TOMATOES: *This large-scale tomato grower needed to rotate his crops to reduce disease and pests by growing non-host plants before a biological program was started. The top left photo shows the problems they had with the tomatoes before adding biology. The top right photo shows the healthy tomato plants after the biological program was implemented. Rotation of fields, or letting fields rest without producing is no longer necessary since the biology has been added. (Photo courtesy of Soil Foodweb Inc.)*

5. ROOTS INTERFACE BIOLOGY AND PLANTS

Roots prevent plants from falling over and help plants to be firmly planted in the soil or the growing media. Lateral roots take up nutrients by simple diffusion. Roots do not make enzymes to break down organic matter and therefore need the help of soil-biology to cycle nutrients from organic matter into plant-available forms. Depending on the type of plant, roots release 10-45% of the energy they fix, as photosynthate. This is released from their roots in the form of simple sugars, proteins and carbohydrates, which are called exudates. These exudates released from the roots are energy/food for the microorganism populations around the roots. The plant, therefore, grows a beneficial microorganism shield right around the root (rhizosphere). In this fashion, non-beneficial organisms (called pathogens) have no food left to grow on, and are simply outcompeted.

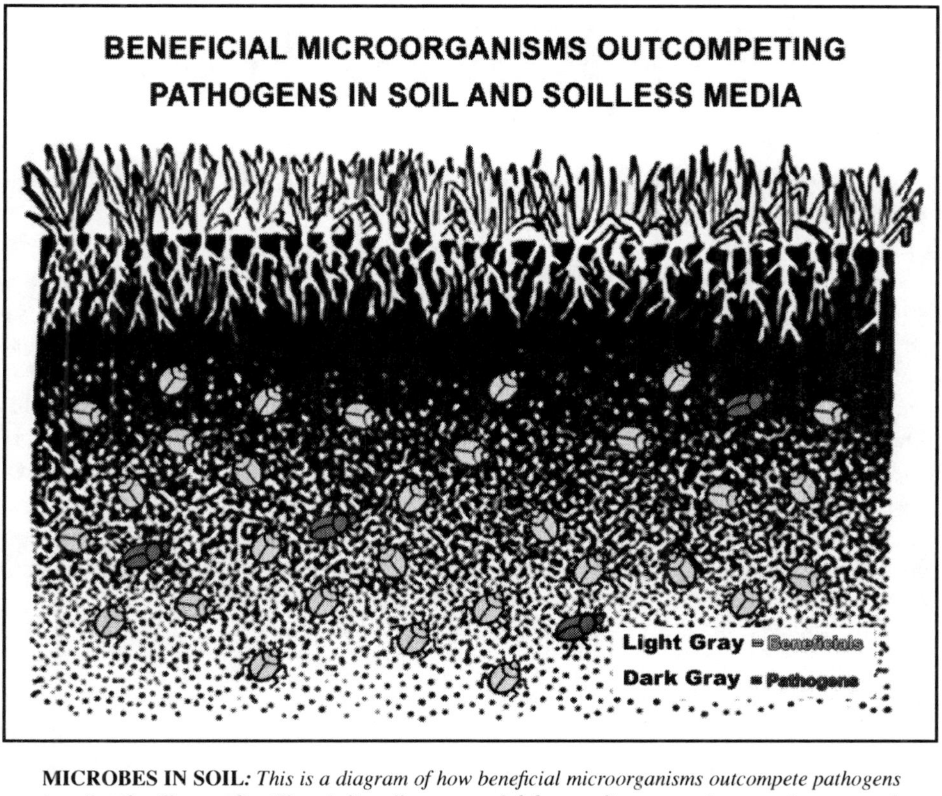

MICROBES IN SOIL: *This is a diagram of how beneficial microorganisms outcompete pathogens in soil and soilless media. There is literally no space left for any disease-causing organisms to reach the plant roots and cause disease. (Diagram courtesy of Nature Technologies International, LLC).*

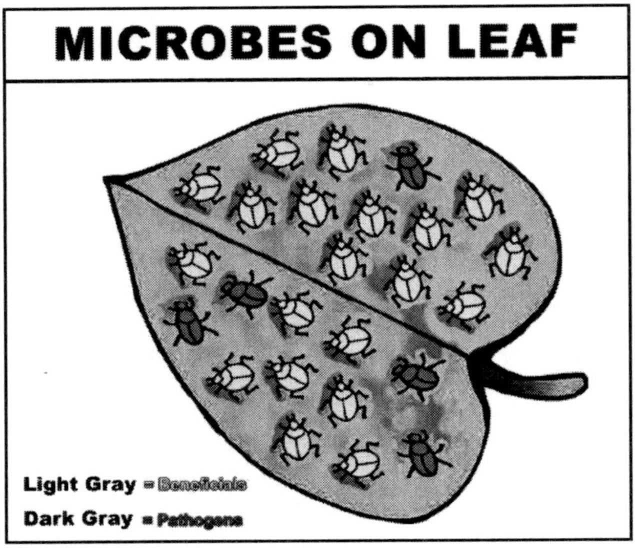

There is literally no place left for any disease-causing organisms

MICROBES ON LEAVES: *This is a diagram of microorganisms covering the surface of a leaf. There is literally no space left for any disease-causing organisms to reach the plant surface and cause disease. (Diagram courtesy of Nature Technologies International, LLC).*

Balance of Beneficial Microorganisms Essential

If the environment in soil or soilless media is out of balance, it means that conditions are present for non-beneficial microbes to grow and potentially harm plants. Without enough good microbes present to compete with, inhibit, and consume the bad ones, the balanced community cannot be maintained. Thus a broad diversity of functions is required of some set of microbes, at all times to perform all the jobs the plant requires, such as: 1) Protection from diseases and pests, 2) Building soil or soilless media structure, and 3) Retaining nutrients in the soil or media. Even small changes in the habitat will awaken some species, while other species of organisms will be put to sleep. Two types of organisms, fungi and bacteria, occupy completely different niches of a habitat (a growing environment). So lots of different species and lots of different individuals of each species are needed for all functions to be performed at all times in all types of habitats and environments.

A balance of biology is also necessary. There are minimum amounts and types of microorganisms that must be maintained; minimum levels of bacteria, fungi, protozoa and nematodes needed to provide protection and nutrient cycling. These ratios can change depending on what type of plant is growing. Different plants require different balances of bacteria or fungal biomass. Diversity of species is needed, no matter what type of plant you are growing.

RENOVATED ANAEROBIC LANDSCAPE: *Before renovation of anaerobic landscape created by using poor quality soils and no drainage, the top two right images depict a classic case of a tree planted in these extreme anaerobic conditions, showing no hair root growth years after planting. (There is water dripping from the tree ball). The high silt and clay content of the soils in addition to overwatering, prevented oxygen movement and created a disease prone environment. By importing a good balanced topsoil (EssentialSoil) and amending of some existing soils with good compost (good biology), aggregates and compost tea, the native area is now thriving within the first year after construction. (Photos courtesy of the Hendrikus Group)*

HYDROPONIC USERS ADDING BIOLOGY: *Several hydroponics users in Northern California have been adding beneficial biology to their systems in the form of compost tea. One user said, "Nothing has ever been so beneficial in so many different ways -- growth, reducing powder molds, reducing bug issues without compromising quality, nutrients or the environment. All the growers I know using compost tea really love it because it increases production." Quality compost, with the proper life is critical, as is the maintenance of aerobic conditions. (Photo courtesy of Nature Technologies International, LLC).*

6. TYPES OF MICROORGANISMS

Beneficial Aerobic Microorganisms

Beneficial aerobic microorganisms require food, air, and water to survive in soil and hydroponic systems. Dissolved oxygen levels must be maintained above 6 ppm (6 mg of oxygen per Liter) for the aerobic organisms to live.

Anaerobic Microorganisms

There are two general types of anaerobic organisms: facultative anaerobes and true anaerobes. Facultative anaerobes live in reduced oxygen conditions, while true anaerobes live in essentially oxygen free, or very low oxygen concentration conditions. Anaerobic microorganisms generally include pathogenic types of organisms that cause diseases. Anaerobic organisms produce metabolites (waste products) that are volatile organic acids (valeric acid, butyric acid, phenols) that often smell like rotten eggs, vinegar, sour milk, ammonia, sulfur or vomit. High nitrate levels, in general, also help select for the growth of disease-causing organisms.

PLANTS IN THE FIELD AND IN POTS CONTAIN BENEFICIAL MICROBES: *The above photos depict some of the many types of plant growing systems and crops which contain beneficial and balanced sets of bacteria, fungi, protozoa, nematodes and microarthropods. Earthworms and insects can also be found in both types of these healthy systems. (Photos courtesy of Nature Technologies, International, LLC.)*

7. TYPES OF BENEFICIAL AEROBIC MICROORGANISMS

Soil and soilless hydroponic systems can contain bacteria, fungi, protozoa, nematodes and microarthropods. Earthworms and insects can also be found in both types of these systems.

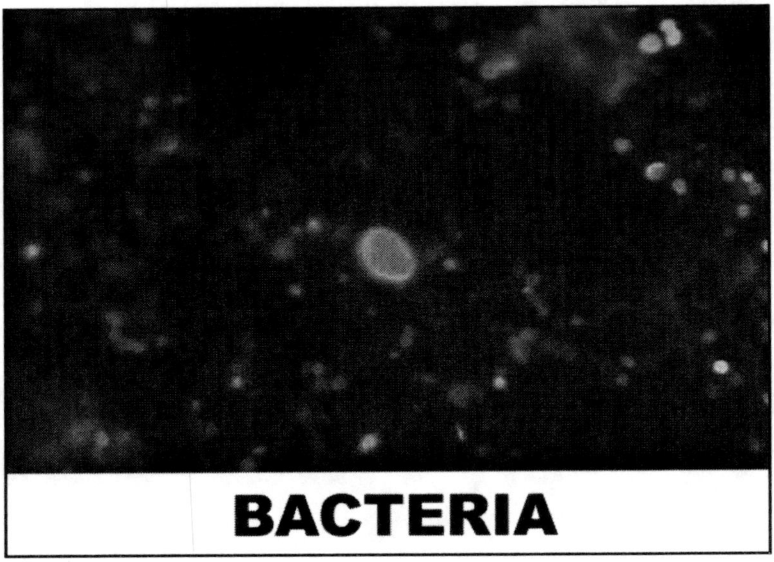

BACTERIA: *The brighter round to rod shaped objects in the photo are bacteria that produce layers of sticky glue around their cells to help keep them attached to surfaces, especially surfaces which produce their food resources. (Photo courtesy of Soil Foodweb, Inc.)*

Bacteria

There are beneficial and non-beneficial bacteria. Bacteria are fed directly by the plants and from organic material added to soil and hydroponic systems. Bacteria do not directly release the nutrients they take up from organic matter all by themselves. Instead they retain nutrients (N, P, S, Ca, Fe, etc.) in their biomass. Protozoa and nematodes are needed to complete the nutrient cycling process. Without all biological components in the plant growing medium, nutrient cycling will not occur.

Bacteria can float free in water without attaching to organic matter or roots, but they preferentially cluster around roots. Bacteria produce glues to attach themselves to surfaces -- leaf surfaces, root surfaces, soil and soilless media surfaces, or to any solid they encounter. The bacteria do better if they glue themselves to surfaces where there is food. Bacteria also glue particles of organic matter together which builds and maintains soil and soilless media micro-aggregate structure. Passageways for air, water and roots to move into any medium are improved through these bacterial processes

Through observation using microelectrodes, the glue layer of aerobic bacteria has been found to always be greater than a pH of 7. Anaerobic bacteria produce acidic conditions (pH less than 7) around their cells. Very different effects on pH occur when a medium lacks oxygen.

Bacteria immobilize nutrients in their biomass. Not only N, but also every nutrient required to keep a cell alive is immobilized in bacterial biomass. The C:N ratio of bacteria is 5:1, the narrowest of any organism living on this planet. It is clear why bacteria will not release N in a mineral form because they have to keep all the nitrogen they obtain in order to survive. Indeed, luxury consumption of N by bacteria is a well-known phenomenon, where the "excess" N is stored in vacuoles inside the bacteria instead of releasing N to the environment. Similar storage capacities have been shown for other nutrients such as P, S, and K in bacteria.

Bacteria also decompose toxic materials. Aerobic bacteria can decompose the toxic materials made by anaerobic bacteria. So, after an anaerobic episode, healthy soil and growing medium conditions can be recovered by allowing aerobic organisms to resuscitate the soil or system.

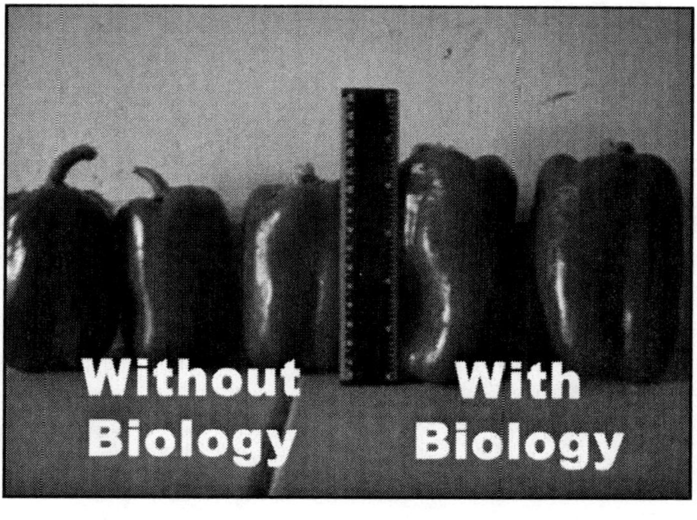

PEPPERS: *Peppers on the right came from plots using actively aerated compost tea. Peppers on the left had no tea applied and used conventional, chemical approaches. (Photo courtesy of Soil Foodweb Inc.)*

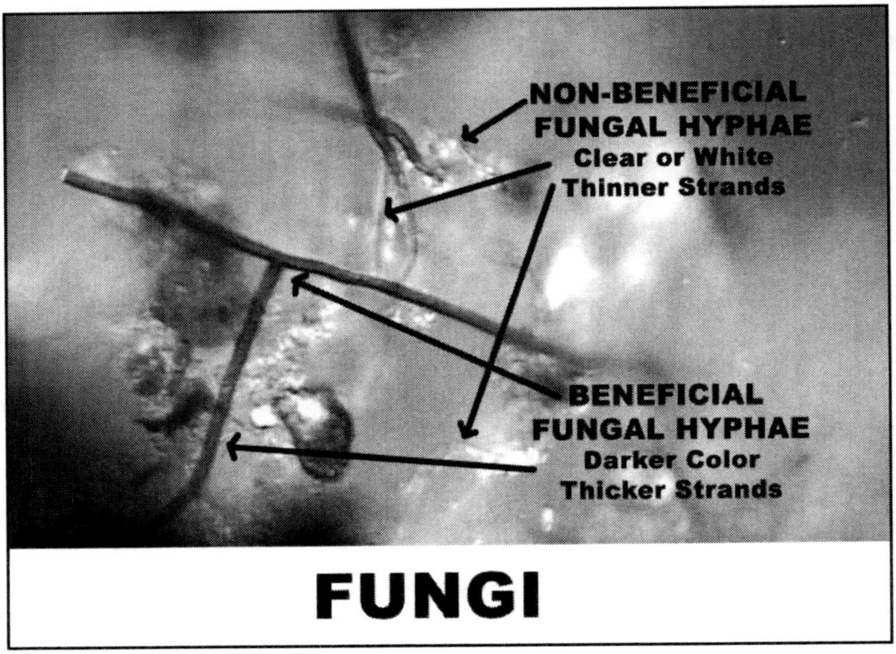

FUNGI: *The dark brown, thick threads in this photo are the hyphae of beneficial fungi. The clear or white, thinner threads are of possibly less beneficial fungi. Beneficial fungi tend to be wider in diameter, colored and have regular cross walls in their hyphae. (Photo courtesy of Soil Foodweb, Inc.)*

Fungi

Fungi grow in long threads that are called hypha (the plural is hyphae). Fungi use those threads to bind themselves to leaf, plant and root surfaces, as well as binding organic debris, soil and soilless media particles together into macroaggregates. Fungi do best when attached to something.

Fungi also work to suppress disease-causing organisms, mainly because beneficial fungi work to improve soil or soilless media structure. When soil or soilless media structure is maintained or improves, water and air can move normally in the soil or soilless media alleviating anaerobic conditions, thus reducing disease-causing organism activity.

Fungi retain nutrients in their biomass much like bacteria, though at a lower concentration relative to carbon. The C:N ratio of fungi is only 20:1; not as concentrated as bacteria. Although there is less holding capacity for nutrients in fungi, these organisms sequester an amazing amount of carbon in soil or soilless media. If the goal is to put carbon back into the soil or system (as discussed on page 13), fungal growth should be selected, since fungi are better than any other organism at tying up carbon and keeping it in an organic form. Indeed, fungi make the most long term, recalcitrant humic acid from complex carbon compounds.

Fungi are also noted for their outstanding ability to decompose certain toxins, especially long chain hydrocarbons and recalcitrant carbon molecules. Fungi are the major decomposers of most complex carbon compounds such as humic acids.

There are beneficial and non-beneficial (disease-causing) pathogenic fungi. Typically disease-causing fungi fall in the category of the Oomycetes, which are hyaline (clear), do not have consistent cross walls (septa), and are narrow in diameter. High nitrate, high ammonium and lack of oxygen typically set the conditions in soil and soilless media for growth of non-beneficial species of fungi.

TURF GRASS STRUCTURE: *Before biology was used, the chemically maintained turf (left side, top and bottom photo of playing field in England) had the worst reputation in the league. Grass was sparse, the soil was very compacted and smelled of ammonia and vinegar. Six months after the biology (in the form of compost tea) and foods were added, the results are shown on the right . No one complains about the club's fields anymore. (Photo courtesy of Michael Harrington, Newby Rugby Club).*

Typical fungi (some of which may cause disease) found in soil and soilless media, fall in the Ascomycetes category. These are wider diameter in general than the Oomycetes, and typically have more color and uniform septa. The most beneficial fungi have wide diameters, consistent septas, and dark colored hypha and are usually found in the last category of fungi, called Basidiomycetes. In general, the beneficials all have dark coloration to their hyphae. There are a very few disease-causing fungi in this last category.

Fungi are the major holders of calcium that is added to soil or soilless media. Plants require calcium (Ca) in a soluble form in order to take it up. Fungi retain calcium on their surfaces as a first step in producing soluble, plant-available calcium (see photo on page 45). Like bacteria, fungi must be eaten by nematodes or microarthropods to release the nutrients inside the fungi and turn them into plant-available forms. When fungi are eaten by fungal feeding nematodes or microarthropods, a chelated form of calcium is released.

Yeasts are a type of fungi that grow predominately in reduced oxygen conditions. As a result of anaerobic metabolic biochemical pathways, they produce fermentative materials such as alcohols. Even tiny concentrations of alcohols (1 ppm, or 1 microgram of alcohol per liter) will kill plants by dissolving the cell wall structure.

LETTUCE: *Alan Magan of Chesapeake Garden in Maryland adds biology in the form of aerobic compost teas. "We spray compost tea on our soilless mix, then we plant lettuce in our greenhouses. Every time we cut the lettuce, we spray the beds again with about 10 parts water and 1 part tea dilution rate. In the summer of 2005 we did a trial on 250 potted sweet peppers. Using the compost tea made a dramatic difference in the fruiting color and growth." (Photo courtesy of Chesapeake Gardens).*

ROOT DISEASE CONTROL: *These roots colonized with vesicular-arbuscular mycorrhizal fungi are protected from disease pathogens. Benefits of VAM fungi: 1) May physically prevent the disease from reaching the root, 2) May utilize the food that the disease organisms need to germinate and grow, 3) May make the inhibitory compounds that prevent disease organisms from growing, or 4) May improve the nutrition of the plant by bringing mineral nutrients from the soil to the plant. No matter the precise mechanism, colonization by mycorrhizal fungi is beneficial by reducing plant stress for most of our agriculturally important plants. (Photo courtesy of Soil Foodweb, Inc.).*

Mycorrhizal Fungi

Mycorrhizal fungi are a type of fungi that enter into a symbiotic relationship with plants. The term *"mycorrhizae"* means the structure built by the root-fungus association. The fungus solubilizes mineral nutrients from sand, silt and clay, and from organic matter. It then transfers the nutrients to the plant in exchange for sugars the plant makes. In this way both organisms "win", by surviving and growing in places neither would manage without their symbiont. *Mycorrhizal fungi* are valuable because they colonize roots and prevent diseases (pathogenic fungi) from being able to establish on those roots.

Pathogenic Fungi

Pathogenic fungi use living organic matter as a source of nutrients and cause the death of their host plant. They cause disease. These are not desirable in any growing system.

Saprophytic Fungi

The largest numbers of fungi are those that use non-living organic matter as their source of food. These are called *Saprophytic fungi* and include the beneficials that help suppress, compete with, and inhibit disease-causing organisms.

Protozoa eat bacteria and release nutrients available for plant uptake, such as, nitrogen released as NH_4 (ammonium).

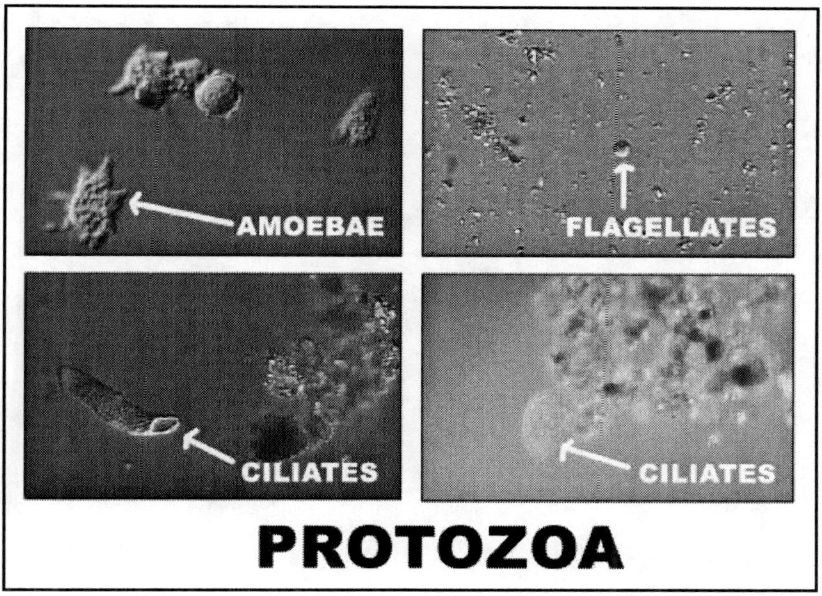

PROTOZOA: *Different types of protozoa – amoebae, flagellates, and ciliates all consume bacteria, which have a much narrower C:N ratio than any of the protozoa. As a result, N is released in a form that is plant-available. Other nutrients are released as well, all of which are in a plant-available form. (Photo courtesy of Soil Foodweb Inc.)*

Protozoa

Protozoa include *flagellates, amoebae,* and *ciliates.* Their major function is to make nutrients plant-available and maintain and build structure. Protozoa primarily eat bacteria, but also eat other protozoa and sometimes fungi. Protozoa release nutrients as waste products that stimulate the growth of more bacteria, fungi and plants. Bacteria and fungi retain (immobilize) nutrients in their biomass. Protozoa eat bacteria and release nutrients available for plant uptake[7], such as nitrogen released as NH_4 (ammonium). Other organisms also eat protozoa.

Flagellates – These are generally round to pear shape, strict aerobes with long "tails" called flagella.

Amoebae --- These are also strictly aerobic organisms that "ooze" across surfaces.

Ciliates ----- These graze bacteria, but tolerate reduced oxygen conditions. Therefore high numbers of ciliates indicate conditions where the strict aerobes are not present.

LANDSCAPE AND GREENHOUSE FLOWERS: *These flowers were all started in a greenhouse in Northern California with biological/organic products. Biology in the form of compost tea was added with great success. Flowers were then planted in the landscape, with compost tea added while transplanting. (Photo courtesy of Nature Technologies International, LLC.).*

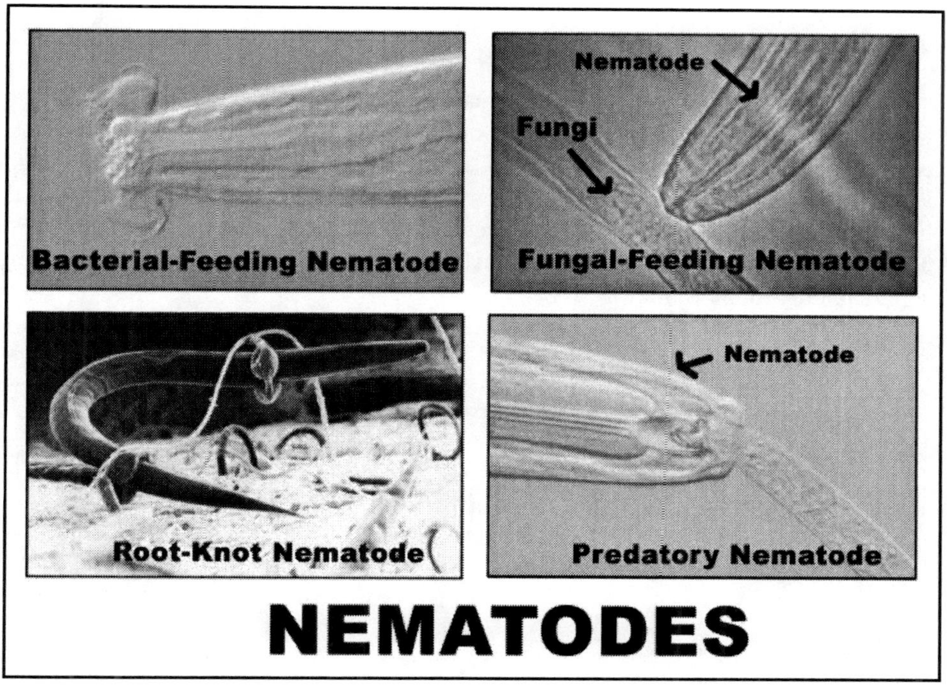

NEMATODES: *Different types of nematodes—bacterial-feeding nematodes in the upper left hand corner, fungal feeding nematodes in the upper right hand corner, root-knot nematodes in the bottom left hand corner, and a predatory nematode eating an organism in the bottom right hand corner. Root-feeders need to be selected against, and the proper biology will do that. The beneficial nematodes will do a great deal to remove and prevent root-feeders from finding roots. (Photo courtesy of Soil Foodweb Inc.)*

Nematodes

Plant-feeding nematodes consume roots and can significantly harm plant production. Conversely, beneficial nematodes work like protozoa to make nutrients available to plants. Nutrients such as nitrogen (N), phosphorus (P), sulfur (S), etc. (originally retained by bacteria and fungi) are processed in the nematode's metabolic system, and then, released in a plant-available form as part of their waste. Certain nematode species feed on bacteria and other species feed on fungi. Beneficial nematodes build structure and also stimulate prey groups. Predatory nematodes consume other nematodes and can keep root-feeding nematodes under control.

Microarthropods

Microarthropods make nutrients available to plants, stimulate prey, and build structure in soil or soilless media, whether rock wool, coir, peat, or water. Microarthropods also carry other microorganisms through the planting medium, from older leaves to younger leaves, and from one root to another. Microarthropods are not normally part of water-based hydroponics systems, but they do occur in soil and soilless media, clay, vermiculite and coir, under certain environmental conditions. If anaerobic conditions are present, certain microarthropods may eat roots.

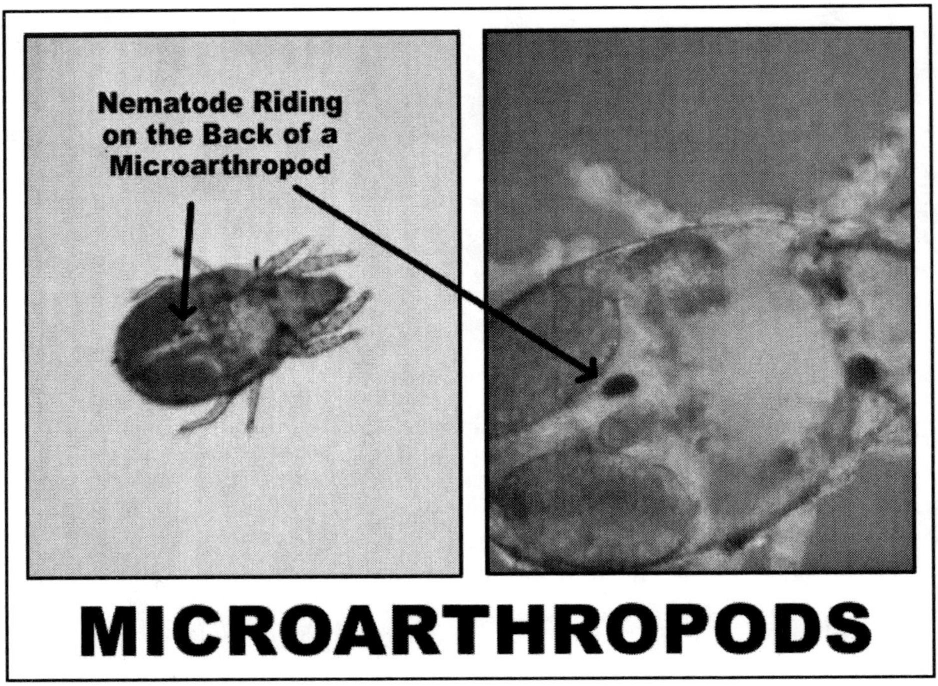

MICROARTHROPODS: *Microarthropods can carry nematodes, protozoa, bacteria and fungi from place to place, serving as taxicabs for beneficial organisms. Microarthropods also feed on fungi, or on other arthropods, and release nutrients in plant-available forms. (Photo courtesy of Soil Foodweb, Inc.)*

RENOVATING ANAEROBIC SOILS: *Standing water demonstrates severe compaction problems at lower depths. What appeared as mere sand, was, in fact, sand with a high percent of silt that easily compacted. Surface drainage conditions were also extremely poor. Die back and unhealthy plants plagued this garden. Existing soils were amended with balanced topsoil (EssentialSoil) and compost. Regular compost tea inoculation achieved a high level of healthy and sustainable growth within the first year. (Photo courtesy the Hendrikus Group)*

8. ENVIRONMENTS FOR MAINTAINING MICROORGANISMS

Microorganisms are like tiny people – just like humans, they need food, air, and water to survive. Different species and communities of microorganisms thrive in different temperatures and environmental conditions. In general, microorganisms grow faster in warmer conditions, with a few exceptions, such as fungi, which grow better in cooler conditions.

Aerobic bacteria tend to make their environment more alkaline (pH above 7), while fungi tend to create more acidic conditions (pH below 7). Although there is no absolute demarcation of aerobic to anaerobic, in general, aerobic organisms stop functioning below an oxygen concentration of 6 ppm. When dissolved oxygen levels are below 6 mg of oxygen per liter, anaerobic organisms will be growing and releasing phytotoxic compounds, such as alcohol and formaldehyde. Many organic acids produced under anaerobic conditions have a pH of 2, such as acetic acid and butyric acid. Anaerobic organisms may not actually drop the pH to 2 because of the mix of other materials present, but pH will certainly be lower in reduced oxygen environments indicating these conditions.

TURF GRASS: *"The top photo is turf grass before treatments were started. The soil was a grade three soil, as described by the USDA in their soil map book for our county. The middle photo is after six months of treatment, and the bottom photo is after 12 months. Over the 12 month period, the turf received about one quarter of the amount of Nitrogen of what ChemLawn would use per year -- and no herbicides." Alan Magan, Chesapeake Gardens, Maryland (Photo courtesy of Chesapeake Gardens).*

FLATS IN GREENHOUSE: *Flats of flowers and vegetables in a greenhouse in Northern California in a completely organic/biological system maintained with compost tea. There were no disease or pest problems and great roots developed. No pesticides or synthetic nutrients were used. (Photo courtesy of Nature Technologies International, LLC.).*

Hydroponic media are designed to have excellent air passageways and structure from the beginning. When biology is added to plant growing environments, they build and maintain the air passageways and structure in soil or soilless media. Some growing systems, however, develop biofilms. Biofilms are layers of microorganisms, and their waste, that get glued to soil particles, aeration tubes, recirculation tubes, tanks, pots and growing media clogging air passageways and soil or media structure. If too much food is added, and not enough predators are present to deal with the exploding populations of microbes (multiplying from too much food), layers of bacteria can grow quite thick and clog the air passageways. These thick layers do not allow oxygen to penetrate, thus creating an anaerobic environment for pathogens and non-beneficial organisms to thrive. We call these thick layers "biofilms."

Predators (protozoa and nematodes) are required to keep the numbers of bacteria and fungi in balance to prevent these biofilms from forming. Minimal amounts of bacterial and fungal foods should be added to any soil, or soilless solution water, to limit population growth and biofilms. Let the plants feed the microbes for their particular growing conditions.

OVERUSE OF PESTICIDES AND INORGANIC SALTS: *Overuse of pesticides and inorganic salts ("synthetic fertilizers") has reduced the normal sets of microflora, leaving conditions open for pathogens to survive and grow, without competition, in conditions where they would not normally be a problem. (Photo courtesy of Nature Technologies International, LLC.).*

Dissolved Oxygen Issues

Most beneficial organisms are aerobic; only a few beneficials thrive and are competitive in reduced oxygen environments. Reduced oxygen conditions select for the growth of organisms whose waste products, such as formaldehyde, phenol, and many organic acids, will harm plants. Quite often, then, we consider these organisms to be pathogens. Aerobic conditions will suppress the growth of these pathogens, so maintaining dissolved oxygen levels above 6 mg/L prevents these problems from occurring. Most, but not all, plant pathogenic and human disease-causing organisms grow best in reduced oxygen conditions. However, these pathogens can often grow in aerobic conditions if the normal aerobic organisms are not present. Overuse of pesticides and inorganic salts ("synthetic fertilizers") has reduced the normal sets of microflora, leaving conditions open for these pathogens to survive and grow without competition in conditions where they would normally not be a problem.

Consider that:

1) Oxygen in the air decreases as elevation increases
2) Organisms grow faster as temperature increases
3) Organisms grow faster as more food is added to a system

As elevation or temperature increases less food can be added in liquid mediums. As elevation or temperature increases, microbial growth will result in anaerobic conditions more rapidly. The likelihood of growing pathogens increases if pathogens were present in the medium at the beginning. Thus fewer foods should be added to any growing medium as elevation or temperature increases.

WETLANDS: *Wetland plants thrive best in balanced soils, as does any plant. The nutritional and microbial balance differs from terrestrial plants, but remains important. These images show constructed wetlands and ponds that feature deep amended soil layers installed at the ponds edge. Soil was designed to promote root growth and retain the thin, but important, upper aerobic layer of soil typical of healthy wetlands. A local wetlands and ecology expert noted that he saw growth occur "in one year that normally takes five years to accomplish." (Photo courtesy the Hendrikus Group)*

pH Issues

Chemical based agriculture takes the point of view that if nutrients are in solution, they must be available to plants and solubility of nutrients is pH dependent. Different nutrients are more or less soluble at different pH levels, but the optimal pH for most nutrients for plant growth is at pH 6.5 – 6.8. Therefore the current paradigm says all plants must be grown at this pH.

In fact there is little relationship between the concentration of nutrients in water and what plants take up. Nutrients can be high in solution without much being taken up by the plants, or they can be low in solution and the plant gets everything it needs. There usually is little correlation between plant uptake and soluble nutrient concentrations. Given these observations, it would seem that trying to control plant growth by maintaining optimal pH falls short of benefiting plant production.

NUTRIENT AVAILABILITY: *Chemical based agriculture takes the point-of-view that if nutrients are in solution, then they must be available to plants...in fact, there is little relationship between the concentration of nutrients in water and what plants take up. (Photo courtesy of Nature Technologies, International, LLC.)*

Plants take up chelated nutrients more rapidly than non-chelated nutrients. Once a nutrient is chelated, pH becomes unimportant in determining solubility. Biological processes typically perform chelation. If biology is present to chelate nutrients, then pH is of no consequence.

Another factor to consider is the proper form of nutrients for different plants. The case of the proper form of N for different plants probably exemplifies this situation best. Agricultural studies have shown that addition of nitrate fertilizer to common AG crops increases plant production. The conclusion was that addition of nitrate fertilizer would be beneficial for all plants. Except that with time, more fertilizer had to be added with less and less benefit to plant yields. And then it was discovered that perennial plants require ammonium (NH_4) as a preferred N source, not nitrate (NO_3). Many crop plants require a balance of nitrate and ammonium, while some plants do best with just nitrate.

What controls this balance? Most likely the biology. Predator prey interactions between protozoa and bacteria, nematodes and fungi, or microarthropods and fungi, release N as ammonium. In soil, where fungi predominate and maintain soil pH between 5.5 – 7.0, the ammonium stays ammonium. But where soil is bacterial-dominated, and aerobic, then pH is usually greater than 7, and considered alkaline. In these cases N is present as NO_3 because nitrifying bacteria grow and convert NH_4 to NO_3 only in alkaline conditions.

Nitrifying bacteria are not killed by pH lower than 7.0; they just don't function to convert N to NO_3 at low pH. Therefore plants which require NO_3 require pH greater than 7.0, if growing in conditions where inorganic fertilizers are not constantly applied. Even though solubility in H_2O, of N, P, Ca, Fe, etc., may be highest at pH 6.5 – 6.8, the *form* of these nutrients may not be correct for plant uptake.

Given the above discussion where a predominance of aerobic bacterial growth drives soil into alkaline conditions and a predominance of fungal growth maintains pH between 5.5 and 7.0, how can soil achieve a pH below 5.5 which is often observed in soil? As anaerobic bacteria grow, they make organic acids with low pH, which drops soil pH into pH ranges below 5.5. Thus low pH indicates anaerobic microbial growth. Can a soil be lacking oxygen, but still be above a pH of 5.5? If the buffering capacity of the soil has not been neutralized by anaerobic acid production, then pH can be high while anaerobic bacteria and yeasts dominate the soil.

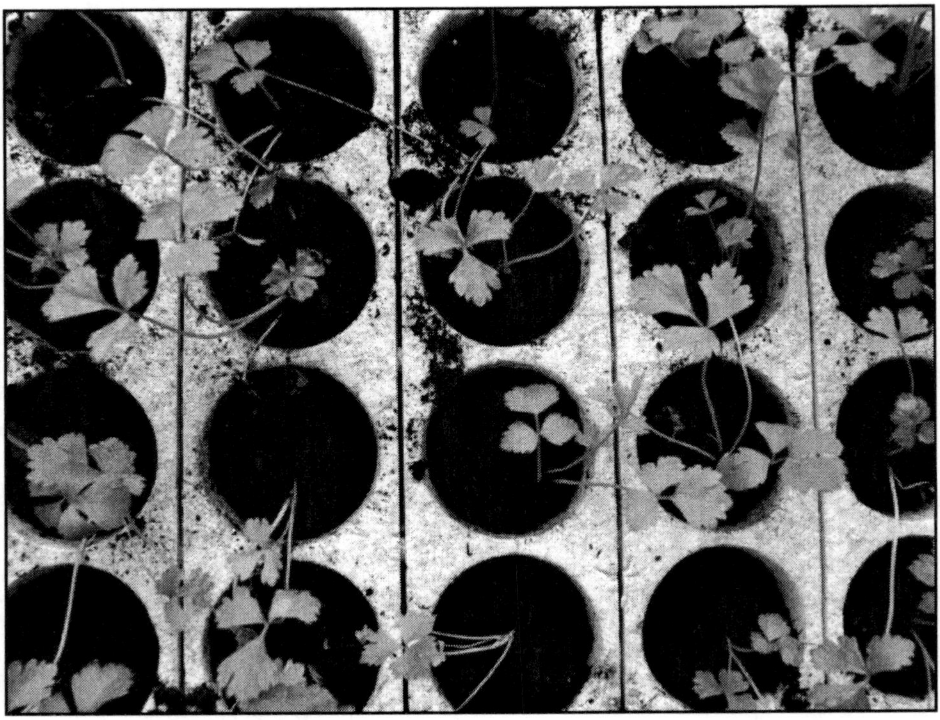

CHELATED NUTRIENTS: *Plants take up chelated nutrients most rapidly. Biological processes typically perform chelation. If biology is present to chelate nutrients, then pH is of no consequence because the biology will control the proper pH for the plant. In healthy systems, the plant is what controls microbial activity (Photo courtesy of Nature Technologies International, LLC.)*

Electrical Conductivity

Electrical Conductivity (EC) indicates the ionic strength of solutions. If EC is 10 and then drops to 9, ionic nutrients were removed from solution, presumably by growing plants. Except that bacteria and fungi can take up nutrients as well. So there is no guarantee that just because EC dropped it is the plant that accessed those nutrients, unless the plant is growing in a sterile medium. In addition, chelated nutrients can not measured by EC, since a chelated nutrient is often electronically balanced. Electrical Conductivity should be maintained between .01 and 1.0 millisiemens (mS) to make sure the solution is the right environment for microorganism populations to thrive. If EC readings are too high, then the salts are too high, and biology may die because it cannot function. If EC readings are too low, then look at the biology and determine whether more biology is needed, change solutions, or add nutrients.

9. MICROBES AND PLANTS FORM A SYMBIOTIC RELATIONSHIP

Microorganisms maintain a symbiotic relationship with plants. The symbiotic relationship that pathogenic microorganisms form with plants harms or kills plants. The symbiotic relationship that beneficial microorganisms form with plants enhances plant growth, health and protects plants from attacks by pathogenic microorganisms. Leaf, stem, flower, root, seed surface, soil particles and soilless media are all places where communities of beneficial microorganism species can live. Each plant selects for different sets of species of organisms for different environmental growing conditions.

Plants Feed Microbes and Microbes Feed Plants

Plants make foods to feed microbes through the process known as photosynthesis. The foods made are sugars, carbohydrates, and proteins (amino acids). Plants release these foods as exudates through their leaves, stems, roots, and blossoms. Microorganisms cluster about the surfaces of plants, waiting to be fed. Organic matter that we add into our plant growing systems also provides nutrients that microorganisms immobilize and retain. Different microorganisms work differently in how they transform those nutrients into a plant-available form. Some bacteria and fungi produce acids that solubilize rock bound P, K, S, Co, B or other nutrients, into forms which these organisms can then take up. Bacteria and fungi have to actually be eaten, themselves, by protozoa and nematodes, who then release those nutrients in a plant-available form as part of their waste materials.

TRANSFORMING NUTRIENTS: *Different microorganisms work differently in how they transform nutrients to get them into a plant-available form. (Photo courtesy of Nature Technologies International, LLC.)*

NUTRIENT BALANCE: *The microbes and the plants work out the nutrient balance between themselves, without any intervention necessary from the outside. (Photo courtesy of Nature Technologies, International, LLC.)*

Fungal and Bacterial-Dominated Environments

Some plants produce more foods that feed bacteria, while others produce more foods that feed fungi. Different plants require different balances of bacteria and fungi because they need different balances of foods and nutrients. If the fungal to bacterial ratios are balanced for any particular growing environment, then the nutrient balance will be maintained by nutrient cycling. The microbes and the plants will work out the nutrient balance between themselves, without any intervention necessary from the outside. However, if nutrients are removed in plant harvest (i.e. corn ears removed, wheat seed removed, apples picked, etc.), then, those nutrients will need to be added back into the system in the form of organic matter (compost, compost tea, rock powders, kelp, etc.)

Bacterial-Dominated Growing Environments

Some plants produce more foods that feed bacteria – so more bacteria grow in that environment. We call this environment bacterial-dominated. Aerobic bacterial glues maintain the pH above 7.0 which creates an alkaline system. When the growing environment is aerobic, bacterial-dominated nutrients are tied up in bacteria (some fungi also). Protozoa eat the bacteria and release the plant-available nutrients. The form of N that is needed in bacterial-dominated soil and media is found in alkaline glues (formed by bacteria) and in nitrifiers. Aerobic bacterial-dominated media maintain an alkaline pH, therefore, nitrate is the dominant pool of inorganic N in these environments.

Fungal-Dominated Growing Environments

Other plants produce more foods that feed fungi – so more fungal populations are maintained. We call this a fungal-dominated environment. Fungi maintain pH levels between 5.5 and 7.0, by producing organic acids in that pH range, so this is considered an acidic system. Nutrients are tied up in fungi (some bacteria also). *Amoebae* (one type of protozoa), nematodes, and microarthropods eat fungi and release the plant-available nutrients. Aerobic fungal-dominated media maintain a pH between 5.5 and 7, and therefore, ammonium is the dominant form of N.

FUNGAL-DOMINATED ENVIRONMENTS: *Grapes require a fungal-dominated environment. The vines produce foods that feed fungi -- so more fungal populations are maintained in that plant growing environment. (Photo courtesy of Nature Technologies International, LLC.)*

The plant selects and feeds those organisms that prevent diseases around its roots, leaves, stems, etc.

Diversity of Microorganism Community Essential

The total number of species, as well as abundance of species, are a measure of diversity of a community of organisms. The number of functional groups of organisms in any plant growing environment is a measure of functional diversity. A wide diversity of species of each functional group is needed because the environment, at any given time, selects for the growth of a very limited set of those species. The environmental conditions will be right for some species of organisms to be working, while others will be asleep. A few to several hundred will match the growth conditions at any moment, so thousands of species of each functional group are needed. The plant selects and feeds those organisms that prevent diseases around its roots, leaves, stems, etc. We do not know the names of all these organisms, but we do know that their numbers in the right quantities work to suppress disease, retain nutrients, and decompose residues. If there are active, functioning, beneficial species for each set of conditions throughout the year, then diseases or pests cannot get established. Diverse sets of organisms will out-compete disease pathogens and pests for food and space, and occupy potential infection sites. For this healthy competition to occur 60-70% of the plant surface (leaf, stem or roots) must be covered with bacteria and 2-5% must be covered with fungi.

Bacteria and Fungi Retain Nutrients

Calcium is retained as crystals on the surface of the fungal hyphae (threads that form the fungi). Other nutrients are also retained in this same manner. Plants cannot get access to nutrients retained by the bacteria or fungi directly; they need protozoa and nematodes to consume them and release the nutrients in a plant-available form. The metabolic processing in protozoa and nematodes transform and chelate these nutrients, which are then released in their waste. The nutrients are then in a plant-available form, ready for plant uptake.

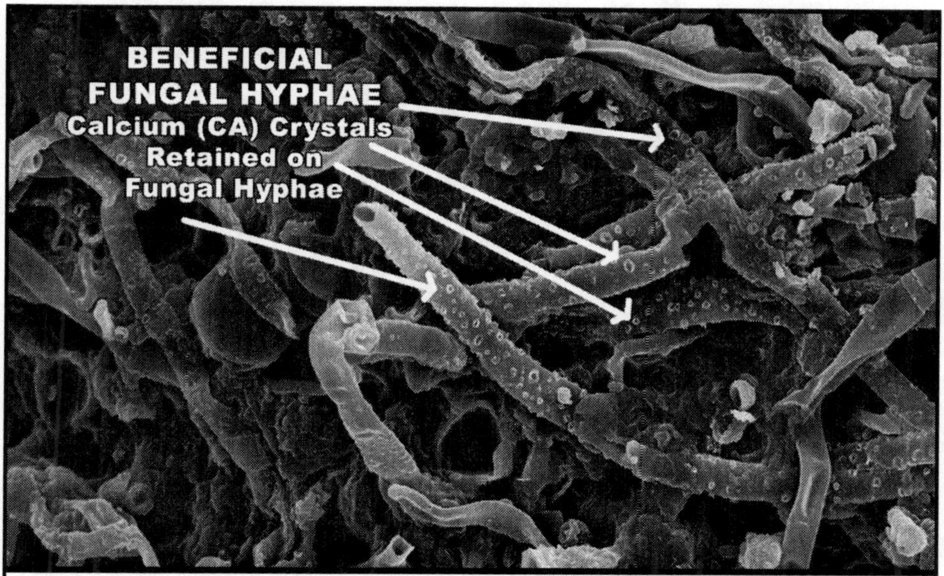

CALCIUM ON FUNGI

CALCIUM ON FUNGI: *Scanning electron micrograph of compost fungi. The crystalline structures on the fungi are calcium oxalate, held and prevented from leaching out of the soil. Mycorrhizal fungi can translocate this calcium directly to the plant. Other fungi will need to be consumed by nematodes or microarthropods to release the calcium in a chelated form for plant uptake. (Photo courtesy of Graham Lancaster and Soil Foodweb Inc.)*

The crystalline structures on the fungi are calcium oxalate

VINEYARD APPLICATIONS OF COMPOST TEA

CLINE CELLARS VINEYARD -- *"With the areas we've been treating with the Actively Aerated Compost Tea solutions for the last two years, tonnage is up by 30 percent this year, where other areas not treated, tonnage is down by 30 percent. With the use of the Actively Aerated Compost Tea Solutions at the rate of 15 gallons per acre, we've noticed that our canopy looks better. I've noticed that the vines are a lot greener than they ever have been. We have no leaf margin burn. We've tried everything in the book, this is the best results we've seen." Tea has been applied to about 400 acres of vineyard and will be applied to a total f 876 acres. Cline Cellars, Ross Hotchkinson, Director of Vineyard Operations, (Photo courtesty of Nature Technologies International, LLC.)*

Protozoa and Nematodes Release Food for Plants

Plants have access to biologically cycled nutrients 24 hours a day, 7 days a week, and can eat whenever they are hungry. The plants do not need to be put on a strict regimented feeding schedule, as inorganic systems require. In an inorganic system, when plants are fed according to a precise feeding schedule, the plant may not need that nutrient at that precise time. The nutrients are then not used and float down the drain or become lost in runoff wastewater.

In an organic system the foods transformed by the microbes into plant-available forms are there all the time. Bacteria and fungi use any excess nutrients not used by the plant, so nutrients are not lost via leaching or run-off. Without excess nutrients available to them, pathogens have no access to high nutrient levels. As a result, many disease-causing organisms cannot grow in organic systems without nutrients in high concentrations.

10. INTEGRATING BIOLOGY INTO PLANT GROWING SYSTEMS

Several soil and soilless growers have success by adding microbes to their existing systems and continuing with their prescribed inorganic nutrient feeding schedules. There are a few guidelines, though, that we can offer to maximize the benefits of those choosing to add biology to an inorganic system, those combining an inorganic and organic approach, or for those using a strictly organic system.

Pumps

Pumps are used to deliver nutrients to irrigation systems, whether for soil drenching or overhead spray applications. Most hydroponic systems utilize air and water pumps to provide circulation of water, nutrients and dissolved oxygen. Water circulation systems can use impeller type pumps that may harm microorganisms when passing through the blades of the pump. Diaphragm pumps may be a better choice when putting microorganism inoculums directly into the reservoirs or irrigation systems. Microbial testing needs to be done with each type of pump to determine which ones may or may not harm the microorganisms.

PUMPS SHOULD NOT HARM BIOLOGY: SPECIALLY MADE SPRAYER AVOIDS DAMAGE TO MICROORGANISMS: *The Sonoma Mountain Institute had a special sprayer made to be able to apply compost tea reaching 50 feet up into the oak tree canopies on their 380 acre property. The sprayer has a 13 HP Honda Motor that delivers the tea at an adjustable psi (50-400) with a diaphragm pump. HPM Sales in Chico, California made the spray rig to very stringent specifications. Sam Sindt, shown in the photo above, is a budding 7 year old Compost Tea Entrepreneur, with his compost tea stand at the end of the driveway for all passerbys to purchase freshly made compost tea. His dad, Mark Sindt, owns a Nature Technologies Compost Tea Production Center in Colfax, CA. For information about the sprayer contact Sonoma Mountain Institute listed in the Resource Section at the back of this book. (Photo courtesy of Sonoma Mountain Institute).*

Checking Levels of Oxygen, pH and Electrical Conductivity

Using meters to check for dissolved oxygen levels in your soil and holding reservoirs is essential. Levels must be maintained above 6 ppm at all times to retain aerobic beneficial microorganism communities. pH is less important to test, since the biology, if adequately maintained, will be able to adjust the pH for whatever plant type is being grown. Therefore measuring pH before plants are added would still be necessary, but relying on pH to determine a feeding schedule is not likely the best choice. Electrical Conductivity (EC) levels will vary according to the types of nutrients used. EC rates can be measured to maintain salt levels between 0.01 and 1.0 mS to make sure that the solution is the right environment for microorganism populations to thrive. Testing needs to be done to determine how much of an inorganic salt-based nutrient can be added before harming the microbiology. For example, 3 ppm copper (Cu) will kill organisms, while 100 ppm of calcium (Ca) can be tolerated by microorganisms before causing harm.

Compost Tea
Foodweb Analysis

Report prepared for:
Sonoma Mountain Institute
Jacob Quinn
4080 Manor Ln
Petaluma, CA 94954-9218 US

jacobquinn@hotmail.com

Report Sent: 05/02/2006
Sample#: 01-102348
Unique ID: T-042706
Plant:
Invoice Number: 0
Sample Received: 04/28/2006

For interpretation of this report please contact:
Local Advisor: or regional lab
Soil Foodweb, Inc
info@soilfoodweb.com
(541) 752-5066
Consulting fees may apply

Organism Biomass Data	Tea Volume (ml)	Active Bacterial (µg/mL)	Total Bacterial (µg/mL)	Active Fungal (µg/mL)	Total Fungal (µg/mL)	Hyphal Diameter (µm)
Results	1	57.0	1984	55.0	57.7	4
Comments		Good	Good	Excellent	Excellent	
Expected Range Low		10	150	2	2	
Expected Range High		150	3000	10	20	

	Protozoa Numbers/g			Total Nematodes #/mL	Percent Mycorrhizal Colonization	
	Flagellates	Amoebae	Ciliates		ENDO	ECTO
Results	57536	42635	**13**	**0.68**	Not Ordered	Not Ordered
Comments	High	High	Low	Low		
Expected Range Low	1000	1000	20	2		
Expected Range High			50	10		

Organism Biomass Ratios	Total Fungal to Total Bacterial	Active to Total Fungal	Active to Total Bacterial	Active Fungal to Active Bacterial	Plant Available N Supply (lbs/acre)
Results	0.03	0.95	**0.03**	0.96	200+
Comments	Good	High	Low	Good	
Expected Range Low	0.01	0.1	0.1	0.9	
Expected Range High	0.1	0.25	0.25	1.1	

Nematodes per Ml of Tea Identification to genus		
Bacterial Feeders		
Bulterius		0.05
Culicularia		0.20
Diplogasteritus		0.02
Rhabditidae		0.40
Fungal/Root Feeders		
Aphelenchoides	Foliar nematode	0.02

728 SW Wake Robin Avenue Corvallis, OR 97333-1612 USA
(541) 752-5066 ; info@soilfoodweb.com
www.soilfoodweb.com

01-102348: Page 1 of 2

MICROBIOLOGICAL TESTING REPORT--Page 1: *Testing is needed to determine specific parameters for each type of system and growth program. With occasional testing by a microbiological lab, you can become fairly precise in your feeding programs and schedules. You can have tests done by a microbiological testing laboratory that provides reports as in the above exampl,e or you can purchase a microscope and take an instruction class on its usage. You will be able to monitor the biology present in your soil or hydroponic system at all stages of growth. (Report courtesy of Soil Foodweb, Inc.)*

Sonoma Mountain Institute
Jacob Quinn
4080 Manor Ln
Petaluma, CA 94954-9218 US

jacobquinn@hotmail.com

Report Sent: 05/02/2006
Sample#: 01-102348
Unique ID: T-042706
Plant:
Invoice Number: 0
Sample Received: 04/28/2006

For interpretation of this report please contact:
Local Advisor: or regional lab
Soil Foodweb, Inc
info@soilfoodweb.com
(541) 752-5066
Consulting fees may apply

Dry Weight:	
Active Bacteria:	Activity in normal range for good compost tea
Total Bacteria:	Bacterial biomass and diversity in expected range; good extraction and growth are indicated
Active Fungi:	Fungal activity above expected levels; fungal biomass will increase as long as nutrients are available
Total Fungi:	Fungal biomass and diversity above typical range for compost tea
Hyphal Diameter:	Excellent, Disease suppressive fungi were extracted
Protozoa:	Aerobic protozoa above desired range. Tea will provide an excellent protozoan inoculum when applied to the soil
Total Nematodes:	Low numbers and diversity. Check compost. Were nematodes present in compost?
Mycorrhizal Col.:	
TF/TB:	Balanced fungal and bacterial biomass. Good inoculum of both fungi and bacteria. Check to make sure both groups survive application system
AF/TF:	Excellent fungal activity, excellent fungal biomass. This tea could beapplied at less than the normal volumes
AB/TB:	Activity adequate, good total bacterial biomass
AF/AB:	Balanced compost tea relative to stated purpose will remained balanced until some disturbance alters that balance
Nitrogen Supply:	6.8 tons of yield possible if all biology is functioning

Interpretation Comments:

24 hour brew in 100gal Nature Technologies brewer from Nature Technologies compost, kit and well water. Arrived in 1L Gatorade bottle, smelled like Gatorade. Notes: Intended crop: SOD infected oaks.
Program sample: send additional email copy of report to whaley11@hotmail.com
Actinobacterial biomass = 0.07 ug/ml. Some true fungal diversity, including sprouting spores, but most hyphae covered in bacteria. Good bacterial diversity.

728 SW Wake Robin Avenue Corvallis, OR 97333-1812 USA
(541) 752-5066 | info@soilfoodweb.com
www.soilfoodweb.com

01-102348, Page 2 of 2

MICROBIOLOGICAL TESTING REPORT--Page 2: *This is the second page of the report shown on page 54. This is the written explanation and interpretation of page 1 of the analysis prepared by the Soil Foodweb microbiological testing laboratory. These are reports of a Compost Tea test sent in by The Sonoma Mountain Institute in Petaluma, CA. (Report courtesy of Soil Foodweb, Inc.)*

Checking the Biology

You can purchase a microscope and take classes from the Sustainable Studies Institute (see the resource section at the back of the book). You will be able to monitor the biology present in your soil or hydroponic system at all stages of growth. With occasional testing by a microbiological lab to double-check your accuracy, you can become fairly precise in your feeding programs and schedules. Several sources have reported amounts that have been successfully applied in different field and controlled environment systems. There are several trials currently under way with a variety of hydroponic systems. We hope to have reports available soon for further determinations and fine tuning. If any biological value on leaf surfaces or in soil is too low, add more compost tea to the reservoir and leaf surfaces. Consider adding a microbiological filter to your recirculating system to maintain biological numbers in your growing environment. Microbiological testing is needed to determine specific parameters for each type of system and growing program. It is recommended that each grower test for their particular growing media and conditions until more experimentation is performed and results can be generalized.

Continuously monitor and make adjustments. More is not necessarily better. Balance is the key.

Examples of Plant Growing Systems

Reservoir Systems -- Deep Water Culture, Ebb and Flow, Nutrient Film

COMPOST: Add a small amount (0.1-1%) of well-aggregated thermal or worm compost or worm castings in the structural material or media.

COMPOST TEA: Compost tea can be added directly in the reservoir water or soilless media at the rate of one gallon of tea for every 50 gallons of reservoir water. Tea can also be applied as a foliar application to feed the plants and provide a protective coating. Coat the leaves with the tea on top and bottom. Dilute the tea as directed to get adequate biology in the system. Be sure to dilute enough to avoid staining the leaves. You can send samples of leaves to a testing laboratory to see if coverage is adequate for biological protection.

ORGANIC NUTRIENTS: Organic nutrients can be added at the rate of 1:200 or as directed by the manufacturer.

RESERVOIR WATER: The water circulating in your reservoir must be aerated to maintain dissolved oxygen levels above 6 mg of oxygen per Liter (or above 18% or 75% oxygen depending on what type of O_2 the probe measures).

ELECTRICAL CONDUCTIVITY: Electrical Conductivity rates can be measured to maintain salt levels between 0.01 and 1.0 mS to make sure the solution is the right environment for microorganism populations to thrive.

MICROBIOLOGICAL TESTING: Bacterial to Fungal ratios can be measured by sending a sample of your reservoir solution to a testing laboratory for microbiological analysis to see if the biology is adequate for nutrient cycling and disease prevention.

PLANT TISSUE ANALYSIS: Plant tissue can be analyzed for nutrient levels, such as soluble nitrogen (N), phosphorus (P), and potassium (K).

BIOFILMS: Watch for biofilms (see page 41) that may develop on reservoir walls and in piping. If low concentrations of nutrients are added (as recommended for organic systems) and high protozoa are established around the roots, then biofilms should not be a problem. Continuously monitor and make your adjustments accordingly. More is not necessarily better. Balance is the key.

Drip Irrigation

The biological solutions must be properly diluted to prevent biofilm formation in the drip lines and emitters. Use 0.1 ml of tea per Liter (1 ounce per 100 gallons) of reservoir water to maintain a safe levels. Emitter openings must be at least 400 micrometers in diameter or larger to allow fungi to be able to pass through the irrigation system. Special care with drip lines and emitters is critically important with foliar applications to avoid clogging and enable fungal colonies to reach the plants.

Special care with drip lines and emitters id critically important with foliar applications to avoid clogging and enable fungal colonies to reach the plants.

Aeroponics

Is it critical to maintain constant water films on the roots in aeroponic systems. By adding the compost teas, contamination and disease are significantly reduced. 0.1 ml of tea per Liter (one ounce for every 100 gallons) of hydroponic reservoir water have successfully been applied in these systems.

Foliar applications of compost tea are used, as well as tea applied directly in the reservoir water.

(Photo courtesy of Nature Technologies International, LLC.)

Sustainable Recycling Nutrient Film Technique for Hydroponics

A completely enclosed sustainable recycling system has been developed that maintains live fish in a separate tank as part of a nutrient film technique hydroponic system. The tank is covered to prevent algal blooms and fish are fed small amounts of leftover harvested plant material from the hydroponic system, along with zooplankton. The effluent waste material from the fish tank then goes through a microbial filter made of pelletized, well structured compost to absorb the excess nitrate and ammonium from the fish. After filtration it is added into the hydroponic system as food for the plants. Foliar applications of compost tea are used, as well as tea applied directly into the reservoir water. All leftover media and residue from plants are composted.

By adding compost tea, contamination and disease are significantly reduced

(Photo courtesy of Nature Technologies International, LLC.)

SOIL ON SLOPES: *Soils for use on slopes can now be designed to accomplish many goals, including preventing erosion without artificial protection, storing storm water, and increasing wildlife habitats. Once biology is properly established, the soil will remain stable on steep grades, allowing growth of deep and healthy roots. High microbial diversity and biomass, along with good balanced chemistry, result in high permeability and a high internal angle of friction, preventing movement. (Photo courtesy the Hendrikus Group)*

11. SOURCES OF BENEFICIAL MICROORGANISMS

Laboratories culture known species of microbes, usually by putting them to sleep, and then bottle them for the commercial market. Aerobic microbes are naturally occurring in high quality worm castings, thermophilic compost, compost teas and organic materials. Scientists can only grow approximately .01% of the species of bacteria in laboratories. 99.99% of the bacterial species are still unknown and unnamed, even though we know they exist. Temperature, oxygen, humidity, food resources and interaction with other organisms determine which species of microorganisms are selected to be active and cultured and which remain dormant and asleep.

Dormant Microbial Products

Most microbes packaged for the commercial market are mainly from this .01% of the known species group. They are individually cultured in laboratories and then put to sleep in a dormant spore form to allow commercial packaging and shipping. The organisms need to be inactive when packaged to prevent bottles from bursting due to waste materials released by active organisms. Phosphoric acid is most often used to induce dormancy, which also reduces the pH to around 3.5. These packaged products have varying degrees of shelf life in stores, depending on manufacturer recommendations, purchase date and usage by consumers. When they are purchased and placed in a growing environment with food, air and water, the organisms wake up and go to work. There is great loss of species diversity in packaged biological products.

Many companies produce bacterial inoculums that contain a few species of bacteria, on up to 80 or more different bacterial species. The rest of the important beneficial bacteria species for full diversity are missing. Since, each set of environmental-plant conditions requires a unique set of active bacterial species, probably many needed beneficial species are missing in these packaged microbial products. Very few beneficial saprophytic fungi are mass manufactured. Some fungal cultures, typically for biocontrol of insects, and spores of mycorrhizal fungi are available commercially. It is recommended that growers test leaf surfaces and root systems to learn whether addition of any type of dormant spores is necessary before adding any inoculum.

In addition to the lack of species diversity, when dormant microbes sold in liquid form are applied to leaves, they will not adhere to the leaf surfaces because they have been put to sleep and are not awake and active. Organisms must be active to produce glues and threads to adhere to plant surfaces. Only a handful of companies have protozoa available in the commercial marketplace. Only a few groups of nematode species are presently commercially marketed. The full spectrum of beneficial nematodes is needed and research is being conducted to make these available commercially. Sources for these inoculums can be found in the Resource List at the end of this book.

ONE OR MORE SPECIES ADDITIONS: *Each species of organism has a narrow set of conditions where it is most competitive. Thus, addition of one or two species will not give protection in all conditions. Diversity is required. In non-varying conditions, single species additions are more likely to be effective. It is always safer to apply many species capable of dealing with a variety of problems (Photo courtesy of Nature Technologies International, LLC.)*

Single Species Inoculums

Applying a single bacterial species to control a specific problems often does not succeed because environmental conditions and plant requirements may not favor the growth or survival of that particular single species. Diversity is essential for the organisms to work together in the right balance. An example of three types of single species inoculum currently commercially available are described below.

Trichoderma

Trichoderma may ward off a certain problem, disease or pest – but it will also compete with other non-aggressive beneficial organisms for food and space, creating the same imbalanced system that exists in chemically treated non-biological systems. In addition, if there is another problem that the *trichoderma* species cannot handle, there are not enough other types of beneficial species left that can effectively treat this new situation. If it is necessary to treat a disease outbreak that a single species inoculum such as *Trichoderma* can handle, we recommend adding those microbes during that disease outbreak only. When the outbreak is under control, add back a diverse set of beneficial organisms that were lost.

Pseudomonads

Pseudomonads are a group of organisms with many different species within their group. Many *pseudomonads* are known for their pesticide decomposition ability, but beware that certain species of *pseudomonads* are disease-causing.

Bacillus

Bacillus species are known for insect repellency and suppression of specific fungal diseases through their production of inhibitory compounds. There are also disease-causing species within the *Bacillus* group. As mentioned, different subspecies perform different functions under different conditions. Only use these single species inoculum if you know what environmental conditions that particular species requires, and if it can perform the function you want under those conditions.

Only use a single species inoculum if you know what environmental conditions that particular species requires, and if it can perform the function you want under those conditions.

(Photo courtesy of Nature Technologies International, LLC.)

Dry Microbial Products

Drying typically causes organisms to go to sleep. If the drying process is too rapid, many species of bacteria, fungi, protozoa and nematodes will be lost. Different companies have different ways of slowing the drying of composts and putting the organisms to sleep, but these are proprietary processes used to assure minimum loss of species and diversity. Some products are being developed to carefully handle and prevent loss of species diversity. (See test results of new research on page 90)

WORMS: *Worms can't digest the plant debris they take in; they eat bacteria, fungi, protozoa and nematodes instead of plant material. (Photo courtesy of Nature Technologies International, LLC.)*

Worm Castings/Compost or Vermicompost

Earthworms ingest whatever they can fit in their mouth, but they can't digest the plant debris they take in. Worms also cannot make the enzymes needed to digest bacteria, fungi, protozoa or nematodes. Instead they grind and compress what they ingest, first in their "crop," much like a chicken, and then in their muscular esophagus. Pressure on the microbes causes the cells to burst open and the worm takes up that "juice". Not all of the microorganisms ingested are actually ground up. The excess of microorganisms ingested that are not ground up, pass into the worm's intestines, an area containing high numbers of extremely beneficial bacteria, fungi and protozoa. The soil or compost organisms surviving to this point then become part of the population of organisms that compete, consume and inhibit newcomers. This highly competitive environment with large numbers of established organisms results in the disappearance of human pathogens, most plant pathogens, eggs of pests, most larval stages of pest organisms and root-feeding nematodes.

The worm cast is actually worm "manure" that looks like tiny balls of encapsulated organic matter containing an outstanding community of bacteria, fungi, and protozoa. Some beneficial nematodes survive passage through the earthworm digestive system. It isn't known why only the beneficials survive this process and not the root-feeding nematodes.

If the worms are mistreated, the processing system can be disrupted and the worms can die. Worms attempting to escape their bin is an indication that their habitat is not sustaining them adequately, resulting in reduced quantities and quality of castings. Because natural processes are not occurring, the resulting compost will not be quality material.

The outside slime layer secreted by worms not only helps them move easily through soil and across surfaces, but also kills human pathogens, most plant pathogens, and the eggs and larval stages of many pests. The outside of the worm often contains this slime layer which also destroys *E.coli.* This was demonstrated when worms were dropped into petri dishes that were literally covered with a surface layer of *E. coli.* Within three minutes, the E. coli had completely disappeared in the precise path of the worm's passage over the plate, as well as anywhere the worms consumed, or any place that was merely touched with their mucilage.

Earthworms are an excellent way to manage "clean-up" of organic material. Worms move through material, turning it and leaving air passageways behind them, inoculating with extremely beneficial bacteria, fungi and protozoa in balances that are in the proper range to grow most vegetables and row crops. There are earthworms that are better adapted to early successional stages, while others are better for forests. There is evidence from Canadian universities that nightcrawlers (Lumbricus) can alter the biology in soil to make that soil inappropriate for growing trees. Pine forests in Canada are being devastated by the worms left by fishermen, since these worms select for bacterial growth rather than fungal growth. Earthworms that are vertical burrowers pull organic matter from the surface of the soil into the soil and thus help to make the soil more bacterially dominated. Given a large enough population of earthworms, this can turn the soil from fungal to bacterial dominated with all the resulting changes in forms and rates of cycling of nutrients.

High quality worm castings contain plant growth hormones and a number of decomposition enzymes. Depending on the foods being fed the worms and the conditions in the worm bins, the bacterial and fungal species growing in the worm's digestive system can release many different plant growth hormones and enzymes, as evidenced by the work of Dr. Clive Edwards at Ohio State University.

The worm cast is by definition the material released from the worm's digestive system and not the material processed simply by touching the worm, or that material which surrounds them. Any material that does not pass through the worm's digestive system that is included in the composted material should be called worm compost. Pure worm "manure" should be called castings. Pure castings are difficult to obtain because the small balls of cast need to be separated from the surrounding material the worm is living in?

Read labels carefully. Ask for information on what the worms are fed, and how long the worms are allowed to work on the compost material. Was the worm compost sieved? Sieving may make the compost material look like it is all small pellets, but sieving will reduce the beneficial fungal biomass in the material. Any mixing reduces fungal biomass, so if fungi are important to your efforts, mix and turn as little as possible.

As with any compost, if the end-product is black in color this indicates that the material became anaerobic (lost sufficient air) some time during the composting process. Too much moisture probably occurred preventing movement of air into the compost, which then selected for human or animal pathogens that require reduced oxygen conditions for growth. Low oxygen habitats result in the growth of organisms which convert inorganic forms of nitrogen (N), sulfur (S) or phosphorus (P) into gases. If compost is black it is an indication of anaerobic conditions where production of ammonia, hydrogen sulfate (rotten egg smell), and possibly phosphine gas has taken place. If N, S and P have been converted to these volatile gases, the fertility of the compost has been lost.

If organic matter has become anaerobic for a significant period of time it should no longer be called compost. Composting is defined as "the oxidative decomposition of organic matter." This means compost must remain aerobic during the composting process. The inside of a small percentage of aggregates might be anaerobic, but the outside of the aggregates (casts) must stay aerobic, so any material released from anaerobic areas will be captured by aerobic organisms and not lost. Thus anaerobic microsites are acceptable, and expected. For example, humans beings are aerobic organisms because we must have adequate oxygen for our normal metabolism to function. Within our bodies we have anaerobic microsites, such as our intestines and muscles. Our bodies carefully control these anaerobic microsites and carefully collect the organic acids produced. while the sum total of the human body remains aerobic, just like compost.

Thermophilic Compost

Aerobic compost is the oxidative (with oxygen) decomposition of a mixture of organic matter. Compost usually contains a great diversity of all the important biology groups – predators such as beneficial nematodes, protozoa and microarthropods, bacteria and fungi. Good aerobic compost should not contain root feeding nematodes, weed seeds, human pathogens or plant pathogens. Compost also contains soluble nutrients.

If the composting process has not been maintained as an aerobic process, anaerobic conditions will prevail and the compost will lack beneficial fungi, beneficial nematodes, flagellates and amoebae (the two aerobic groups of protozoa).

AEROBIC AND ANAEROBIC CONDITIONS: *In anaerobic conditions, only anaerobic bacteria and yeasts abound, making waste products that plug and close up air passageways. When air cannot move into the organic material, the aerobic organisms are killed if the loss of oxygen is rapid. As oxygen is used up, anaerobic organisms grow, making compounds that drop pH, result in volatile forms of N, S and P with loss of these compounds, and kill or put to sleep any aerobic biology. Aerobic compost has many air passageways and good aggregate structure, as shown in the top picture, in comparison with the complete lack of passageways and tunnels in the anaerobic compost. (Photo courtesy of Graham Lancaster and Elaine Ingham, PhD.)*

To obtain quality thermal compost, temperatures must reach at least 131 to 135 degrees Fahrenheit continuously for three entire days throughout the entire pile to kill weed seeds, pathogens and pests. It is microbial growth that causes the increase in temperature. Temperatures must not rapidly rise to greater than 155-160 degrees Fahrenheit because rapid changes in temperature will kill beneficial organisms. Organisms go dormant and are not killed by the high temperature when there is not a rapid rise. If the pile is generating heat there must be microbial growth occurring in order to maintain that temperature. If heat is being generated, the compost cannot be sterile. The question would be -- do the good guys generate temperature or do the bad guys? If the pile is aerobic with a good mix of organisms in the original plant material, it is likely that the beneficial species are generating the heat. If there are indications that the pile went anaerobic, the pile is most likely sterile and the bad guys are generating heat.

The compost pile must be turned whenever it reaches too high a temperature to avoid the potential for the lack of oxygen, loss of N, S, P and the build up acidic conditions from too rapid microbial growth. As mentioned earlier, black compost indicates that anaerobic conditions occurred sometime during the composting process. Ammonia and hydrogen sulfide are both gases produced under anaerobic conditions that leave a black residue in compost. This also means that your fertility has blown off into the atmosphere. If an odor of ammonia is evident, nitrogen is being released as a gas. If an odor of rotten eggs is detected, sulfur is being released as a gas. Charcoal is produced under conditions of high temperature and low oxygen concentration. No biology will survive a compost pile that is converted to charcoal. Beware of black "compost."

Aerobic Compost Teas are a water extract of composted material that is brewed for a given amount of time

<u>Actively Aerated Compost Teas</u>

Aerobic Compost Teas are a water extract of composted material that is brewed for a given amount of time. Composted material should include both thermophilic and worm compost or castings to maximize diversity. Teas contain the active and dormant bacteria, fungi, protozoa and nematodes from the high quality composted materials added to the teas. All the soluble nutrients in these composted materials are also extracted into the compost tea. Bacteria, fungi, protozoa and nematodes are fed while brewing to increase numbers and activity of the beneficial species. The brew must stay aerobic to allow the beneficial species to outcompete any non-beneficial organisms. Dissolved oxygen levels are maintained above 6 ppm to maintain aerobic conditions. Below 6 ppm, we would be encouraging pathogen growth, loss of nutrients, and toxins being released in anaerobic conditions.

Leachates, Extracts and Plant and Manure Teas are Not Compost Tea

Simply running water over compost or worm casting would not produce aerobic compost tea. These are called leachates or extracts, may or may not be aerobic, and contain very few organisms. Leachates or extracts of compost or worm castings may contain nutrients, plant growth hormones and enzymes, but do not contain the biology needed to maintain organic/biological growing systems. Leachates are usually produced without aeration or additional foods.

Plant teas are made by soaking different plant materials, but are not compost tea. Manure teas are made with raw manure, not composted materials and are not considered compost tea. Only aerobic compost tea contains the full complex set of organisms extracted from high quality compost or worm castings. These organisms can improve plant growth.

Quality of Compost Teas

Brewer design, temperature, water, and the quality of the ingredients used in making compost tea all determine the quality of the resulting liquid. Most teas need to be used immediately when removed from their aeration to maintain activity and diversity and to prevent anaerobic conditions. There is one company, though, offering a product in a liquid form with live, active organisms that have a shelf life of over one year. They use a unique packaging system that maintains dissolved oxygen levels, guaranteeing aerobic activity. (See resource list at back of book).

QUALITY OF COMPOST TEAS: *Brewer design, temperature, water and the quality of the ingredients used in making compost tea all determine the quality of the resulting liquid. The photo to the right is of David Sakata, removing the tea after it has brewed for 24 hours. David is owner of Wine Country Flower Farm in Temecula, California, a licensed tea center. (Photo courtesty of Nature Technologies International, LLC.)*

COMPOST TEA HELICOPTER APPLICATION

SIZE OF NOZZLES IMPORTANT: *All 60 nozzles on the boom of the helicopter and the filters were removed to allow the actively aerated compost tea to freely flow without restriction, which often harms the fungal hyphae. Compost tea was applied to 300 acres of native shrub oak land for a restoration project for Sonoma Mountain Institute in Northern California. The pilot is shown in the right photo removing a nozzle with an 1/8 inch diameter opening remaining. When he flew overhead it was like a soft rain as shown in the photo on the left. (Photo courtesy of Nature Technologies International, LLC.)*

12. APPLYING MICROORGANISMS

Compost Tea Application Parameters

Tea must be tested before addition to the application tank, then tested again after passage through pump(s), pipes, tubing, nozzles, etc. Generally, we want to test the surfaces of the plant to which the tea was applied to make sure the organisms survived in adequate levels to adequately cover the surface.

The diameter size of the filter system, drip lines, or nozzles should be above 400 micrometers (mm) to allow passage of fungal strands, bacteria, protozoa and nematodes that cannot pass through an opening smaller than 400 mm.

Droplet sizes for foliar applications should be larger than one millimeter in diameter, for teas to be sprayed in bright direct sunlight. Adequate sprayer openings will allow droplets large enough to provide a protective coating against UV rays. If the sprayer is emitting droplets less than one millimeter, growers should avoid spraying teas between 11:00 a.m. and 2:00 p.m. because UV rays can kill microbes as they fly through the air.

Most sprayers require additional water added to the tea to spray the tea evenly over one acre or more. Check the specifications for your sprayer to determine how much additional water needs to be added. For example, if your sprayer requires 80 gallons to cover one acre and you are applying a foliar application to that one acre, five gallons of compost tea and 75 gallons of good quality aerated water would be required. You should add quality, aerated water without chlorine, chloramines, salts or other preservatives that are microbial inhibitory products. If your water source contains any of these microbial inhibitory products, you must remove them before adding that water to your tea.

SPECIALLY MADE SPRAYER AVVOIDS DAMAGE TO MICROORGANISMS: *The Sonoma Mountain Institute had a special sprayer made (shown on page 53) to be able to apply compost tea reaching 50 feet up into the oak tree canopies on their 300 acre property. They are protecting the trees and restoring the native landscape. The sprayer has a 11 HP Honda Motor that delivers the tea at variable psi with a diaphragm pump. HPM Sales in Chico, CA. made the spray rig to very stringent specifications. For information about the sprayer contact Sonoma Mountain Institute listed in the Resource Section at the back of this book. (Photo courtesy of Sonoma Mountain Institute)*

Compost tea emitted from your sprayer should be tested to ensure that the sprayer pump you are using is not killing organisms. Compare a sample of tea in your sprayer with tea coming out of the nozzle of your sprayer to determine whether organisms are being lost in transition through the pump.

We recommend caution when adding food for microorganisms directly into the sprayer. Once tea is finished, additional foods should be added only after additional water has been added to the tea in the sprayer. Addition of any food will almost invariably result in the tea going anaerobic in the sprayer. If straight tea is being sprayed directly on soil or foliage, it is recommended to have one line for the tea and another line for the additional food.

Leaves and soil should be tested to make sure organisms in tea are surviving on leaf surfaces or in the soil. Take a leaf or soil test before spraying onto leaves or soil and test again two weeks later to ensure the organisms survived their addition to the leaves or soil. With hydroponics it is necessary to make sure organisms are surviving in hydroponic reservoirs.

Compost tea should be brewed at a temperature similar to the environment where you will spray the tea. If soil temperature is 55 degree F, then brew tea at 55 degrees. If the soil temperature is 100 degrees F, then brew tea at 100 degrees. If it is necessary to brew at 50 degrees, the growth of organisms will be slowed down, requiring that you double or triple the time for brewing, typically brewing for 48 or 72 hours, instead of a 24 hour brewing cycle required for higher temperatures. Testing is required to ensure the tea is finished If tea is brewed at 100 degrees, it is necessary to monitor the aeration levels also and match the aeration in your brewer with the oxygen demands of the organisms in the tea.

However you apply the compost tea make sure that the organisms in the tea survive on the leaf surface. Test the leaves to make sure they survive spraying.

(Photos courtesy of Nature Technologies International, LLC.)

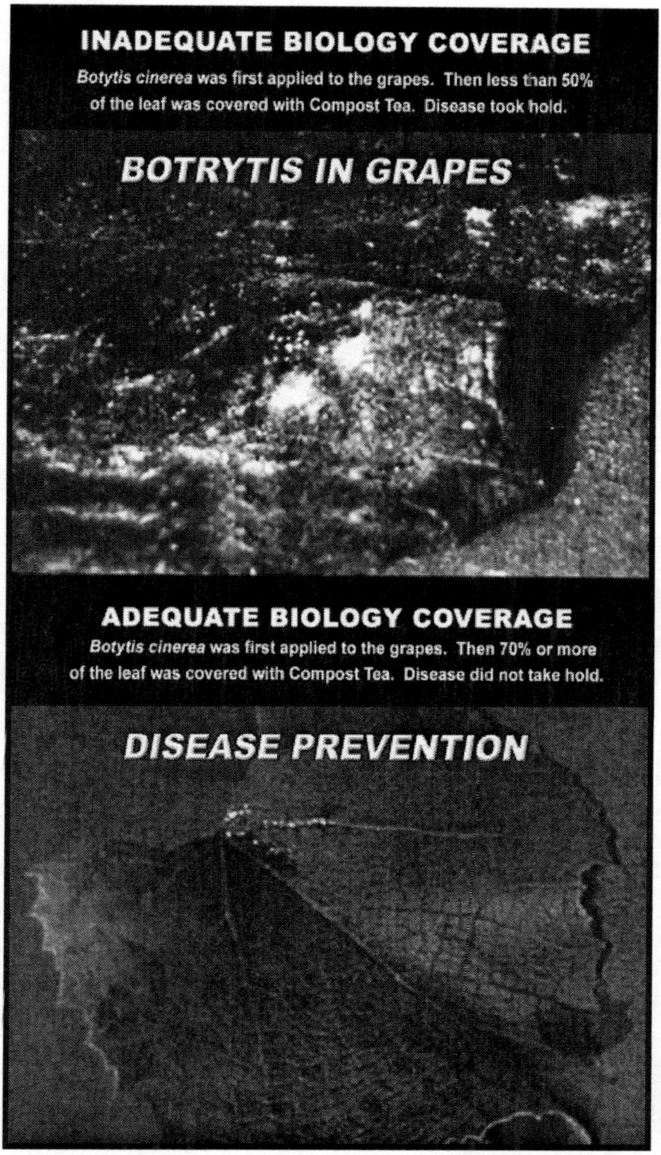

DISEASE PREVENTION: Gray mold was initially applied equally to all plant surfaces. Sections of the vineyard had compost tea applied with 100% coverage, 70% coverage, 50% coverage, 10% coverage or no coverage (water) applied. The grapes with compost tea covering less than 50% of the leaf surface became totally infested with the disease. When coverage was 70% or greater, no disease was observed in any plant in this section of the grape vineyard. Most likely the coverage prevented the disease organism from being able to grow and find plant cells to attack. (Photo courtesy of SARE).

Outside Field Applications of Compost Tea

Apply compost teas at the rate of 20 gallons per acre every Spring and Fall as a soil drench. Apply compost teas as a foliar spray at least three times during each growing season at the rate of five gallons per acre for every five feet of tree canopy. Fertilizer applications can typically be reduced by at least 1/3 each year, as well as watering rates. Use tea for foliar applications as follows: 1st application on the leaves at true leaf stage, 2nd application at the preblossom stage, and 3rd application at the post blossom stage. In leaf applications, the microorganisms raise the CO_2 level around the leaf surface. The amount of CO_2 around the leaf surface determines how much water is in the cells around the openings (called stomata) in the leaf. These are gateways to the inside of the plant. Gatekeepers called guard cells are around these openings. The bacteria grow around the guard cells and keep the stomata open. We are able to see nutrients carried into the plant through this mechanism within 30 minutes after application of teas. When we apply tea directly to the soil or in hydroponic reservoirs it usually takes two to three hours to see plants respond. We can use as little as 1/2 ounce of tea in 50 gallons of water, on up to 64 oz in 50 gallons of water, once a week or every two weeks.

APPLYING COMPOST TEA: *Apply compost teas at the rate of 20 gallons per acre every Spring and Fall as a soil drench. Apply teas as a foliar spray at least three times per growing season at the rate of five gallons per acre for every five feet of tree canopy. (Photos courtesy of Sonoma Mountain Institute).*

APPLICATIONS: Application to foliar surfaces is only required if foliar diseases are normal occurrences. Once soil biology is properly established, the need for foliar applications become minimal. Even in hydroponics, after diseases hiding in the greenhouse are dealt with by applying the proper biology to all surfaces, the need for foliar applications will be minimized *(Photo courtesty of Nature Technologies International, LLC.)*

Seasonal Compost Tea Applications

Seasonal Approach for Annual or Single Season Plants

The seasonal approach for annual or single season plants is as follows:

- **Step one**: Make sure the soil into which the plant will be placed has the correct balance of organisms to select for the desired plant. Make sure the balance is suppressive to weeds. The balance of fungi and bacteria, inorganic to organic N, S, P, etc and micronutrient availability needs to be assessed, and if not balanced correctly, needs to be adjusted.
    - Assess the soil, potting material or hydroponics solution for the full soil food web.
    - Assess soil, potting material or water for exchangeable and soluble soil chemistry.
    - Add any missing nutrients to the compost or water, preferably in a chelated or organic form so the biology will not be harmed.

- Add any missing biology through application of compost or compost tea, whichever is easiest and least expensive.
  - If the soil surface is hard, tilling in compost or applying tea and then tilling may be necessary.
  - Make certain the soil or potting material is moist. Microbes need water just like people, and addition of microbes to a bone dry material will kill them or prevent them from doing their job of protecting and improving conditions for plants.
  - If the compost or tea, fungal to bacterial biomass ratio is not properly balanced, then add additional foods with the compost or tea to help increase the microbial population desired for plants.
  - Put the organisms where they will be able to move to the root and help protect the roots. This may mean injection, tillage, or soil drenching if compaction has been dealt with.
  - Compaction may need to be dealt with before trying to add compost or tea. Organisms of the soil food web build structure in soil AFTER flocculation of the clays has been properly dealt with. Balancing calcium and magnesium, potassium and phosphorus correctly must be dealt with first (see bullet three above).
  - In water add the proper mix of organic nutrients and organisms to the recirculation tank. A microbial filter in which compost is maintained with the right set of microbes can continually remove detrimental organisms and add the beneficial organisms.

- **Step two**: Make sure you have the proper mycorrhizal fungi and biology on the seed surface, or if pesticide treated seed, make sure a high concentration of mycorrhizal spores are present in the soil, potting mix, or in the water below the seed.
  - Soak seed in a compost tea with the right biology, as well as the correct mycorrhizal fungus for your plant.
  - Most vegetables, row crop plants and deciduous trees require VAM (endomycorrhizal fungi). Conifers require ectomycorrhizal fungi, while blueberries and other ericoid plants require ericoid mycorrhizal fungi. Brassicas (mustards) like cauliflower, broccoli or asparagus are not mycorrhizal and there is evidence that becoming mycorrhizal may be detrimental to these plants. Seeing mycorrhizal colonization on a plant may not necessarily mean a benefit to that plant.
  - Once the seed has been soaked in the tea with mycorrhizal fungi, let the seed dry again so planting equipment works correctly. The organisms are in the cracks and crevices of the seed coat and on the seed surface, ready to grow once the seed and new roots begin to release exudates.
  - If the seed has been treated with fungicide, insecticide, antibiotic or any other anti-microbial agent or preservative, the biology must be placed in

the planting row just before the seed is dropped. This requires another tine or injector on the planter from which the compost or tea is released.

- **Step three**: Apply a protective layer of organisms to the plant surface. If adequate microarthropods are present, one application can be adequate. The biology can be maintained by the plant until harmed by pesticide drift, environmental changes, or influx of overwhelming disease problems. Beneficial microarthropods are the taxi-cabs of the soil and above ground plant surfaces. If microarthropods are not present in adequate numbers, then at least three foliar applications of compost tea are required. More applications may be necessary if the soil or the surrounding system is not properly prepared and disease-causing organisms are carried from the soil by wind, rain splash, or insects. Continual replacement of the foliar organisms might be required until soil health is achieved.
    - Compost tea contains actively growing organisms that produce glues to adhere immediately to the leaf surface. Foliar applications REQUIRE actively aerated compost tea. Compost or worm castings do not stick to the leaf surface, nor do dormant or sleeping organisms.
    - Foliar applications of tea may need additional foods to help the organisms get established, stick and maintain activity. Sometimes just addition of the foods is needed to bring up the coverage on leaf surfaces to adequate range.
    - Typically, apply tea through foliar sprays three times per growing season – 1st application on the leaves at the true leaf stage, 2nd application at the preblossom stage, and the 3rd application at the post blossom stage.
    - In leaf applications the excess from the microorganisms causes the CO_2 level to be raised around the leaf surface. The amount of CO_2 around the leaf surface determines how much water is in the cells around the openings (stomata) in the leaf. These are gateways to the inside of the plant. There are gatekeepers, called guard cells, around these openings. The bacteria grow around the guard cells and keep the stomata open. Nutrients can be carried into the plant through this mechanism within 30 minutes after application of teas. When we apply tea directly to the soil or in hydroponic reservoirs, it usually takes two to three hours to see plants respond. We can use as little as 1/2 ounce of tea in 50 gallons of water for up to 64 oz in 50 gallons of water, once a week or every two weeks.

- **Step four**: All residues left after harvest must decompose rapidly so no disease-causing organisms can gain access to foods. Decomposition must be monitored to make sure it is rapid and complete before the next crop is planted. If the residues are not being decomposed (at least half gone in one month with good temperature and moisture), then the proper biology is lacking and must be assessed and returned so that decomposition begins at a normal rate.

HEALTHY ENVIRONMENTS: *The purpose of healthy environments is to grow healthy plants. Health implies an ability to resist disease. If plants obtain nutrients in the proper balance at the proper time. Since these requirements change with stages of plant growth, temperature, moisture and biology in soil, only by having healthy sets of the right biology in the growing medium can plant health be achieved (Photo courtesy of Nature Technologies International, LLC.).*

General Approach to Applying Tea in Perennial Systems

The general approach to applying tea in perennial systems is as follows:

- **Step one**: Make sure the soil or nutrient solution has the correct balance of organisms to select for the desired plant. Make sure the balance is suppressive to weeds. The balance of fungi and bacteria, inorganic to organic N, S, P, etc. and micronutrient availability needs to be assessed and adjusted if not balanced correctly.
    - Assess the soil, potting material or hydroponics solution for the full soil food web.
    - Assess soil, potting material or water for exchangeable and soluble soil chemistry.
    - Add any missing nutrients to the compost or water, preferably in a chelated or organic form so the biology will not be harmed.
    - Add any missing biology through application of compost or compost tea, whichever is easiest and least expensive.

- Compaction may need to be addressed before trying to add compost or tea. If the soil surface is hard, it may be necessary to lightly rake compost into the soil, or apply tea and then rake. As an alternative, remove soil cores down to below the rooting depth, if possible, and refill with a mix of sand and compost (50/50 mix), or spray compost tea (with mycorrhizal spores added) into the holes first and then fill with mix of sand and compost
- It is the organisms of the soil food web which build structure in soil, AFTER flocculation of the clays has been properly dealt with. Therefore, balancing calcium and magnesium, potassium and phosphorus correctly must be dealt with first.
- Make certain the soil or potting material is moist. Microbes need water just like people, and addition to a bone dry material will kill them or prevent them from doing their jobs of protecting and improving conditions for plants.
- If the compost or tea fungal to bacterial biomass ratio is not properly balanced, then add additional foods with the compost or tea to help increase the microbial population desired for plants.
- Put the organisms where they will be able to move to the root and help protect the roots. This may mean injection or soil drenching if compaction has been dealt with.
- In hydroponic systems, add the proper mix of organic nutrients and organisms to the water recirculation tank. A microbial filter, in which compost is maintained with the right set of microbes, can continually remove detrimental organisms and add beneficial ones.

- **Step two**: If planting bare roots, dip the roots in a mix of the proper mycorrhizal fungi and biology.
  - Most vegetables, row crop plants and deciduous trees require VAM (endomycorrhizal fungi). Conifers require ectomycorrhizal fungi, while blueberries and ericoid plants require ericoid mycorrhizal fungi. As far as we know, all perennial plants need mycorrhizal symbionts.

- **Step three**: Apply a protective layer of organisms to the plant surface. If adequate microarthropods are present, one application can be adequate. The biology can be maintained by the plant until harmed by pesticide drift, environmental changes, or influx of overwhelming disease problems. Beneficial microarthropods are the taxi-cabs of the soil and above ground plant surfaces. If microarthropods are not present in adequate numbers, then at least three foliar applications of compost tea are required. More applications may be necessary if the soil or the surrounding system is not properly prepared and disease-causing organisms are carried from the soil by wind, rain splash, or insects. Continual replacement of the organisms through foliar applications might be required until soil health is achieved.

- Compost tea contains actively growing organisms that produce glues to stick immediately to the leaf surface. Foliar applications REQUIRE actively aerated compost tea. Compost doesn't stick to the leaf surface, nor do dormant or sleeping organisms.
- Foliar applications of tea may need additional foods to help the organisms get established, stick and maintain activity. Sometimes just addition of the foods is needed to bring up the coverage on leaf surfaces to adequate range.
- Apply the first application of compost tea about two weeks before budbreak, or at bud-swell. Once leaves or blossoms open, apply monthly, if there are disease considerations later in the season. Especially for blights, shot-hole fungi, scab, scale and mildew, it is critically important to establish the biology on leaf and blossom surfaces BEFORE the non-beneficial fungal spores land. If disease begins to appear, initiate weekly applications of tea for the duration of the fungal infection period. Once the probability of infection is past, return to monthly applications. In the second year, if soil sources of the disease are taken care of, monthly applications will be all that is required; when the disease in the soil and surrounding areas is cleared, only one to two applications a year may be required.

- **Step four**: All residues left after harvest must decompose rapidly so no disease-causing organisms can grow by gaining access to foods. Decomposition must be monitored to make sure it is rapid and complete before the next crop is planted. If the residues are not being decomposed (at least half gone in 1 month with good temperature and moisture), then the proper biology is lacking and must be assessed and returned so that decomposition begins at a normal rate.

Healthy plants put the proper balance of nutrition into their fruits and seeds. Human and animal health requires healthy soil.

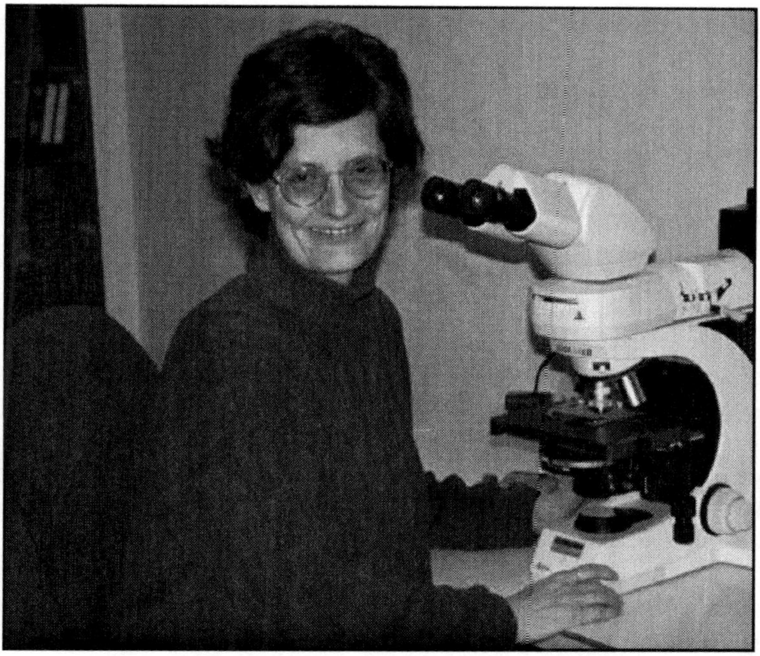

ELAINE R. INGHAM AT THE MICROSCOPE: *At the Soil Foodweb Laboratories in Corvallis, Oregon, testing the microbiological quality of soil, compost, and compost teas is a routine part of each day. All assays are direct microscope assessments, looking at the organisms in their real-world habitats, without trying to get them to grow in laboratory conditions. (Photo courtesy of Soil Foodweb, Inc.)*

12. TESTING FOR BIOLOGICALS

Chemical Analysis

The typical chemistry testing for N-P-K measures only the ionic forms of nutrients. It does not pick up the nutrients in bacteria, fungi, protozoa or nematodes. This is not a good test for what the plant can actually obtain in an organic system. Chelated forms of nutrients are also not always detected by soluble nutrient testing or by Electrical Conductivity (EC) measurements. EC could indicate that nothing is present when plenty of nutrients are being cycled into plant-available forms by the biology. If EC readings are too high, then you need to add organic material to tie up the excess. If EC readings are too low, then check to see if you have adequate biology.

Biological Analysis

Testing requirements are expanded in organic systems. Besides the chemical analysis of N-P-K, there are also tests that measure the amount of microorganisms that perform nutrient cycling in the systems (soil, media, water, on roots and leaves). Additional tests measure the amount of plant-available nutrients (the inorganic forms) that have been transformed by the microbes and are ready for uptake by the plants. Leaf organism assays measure how much biology is present on leaf surfaces to determine if coverage is adequate for disease prevention. Each test helps to determine how much more biology or nutrients need to be added into the soil or soilless system for growing healthy plants.

Types of Microbiological Tests

For microbiological testing send samples from around the roots in soil or controlled environment growing systems. Beneficial or non-beneficial microbes are identified by morphology (physical characteristics) and not by genus or species. We can see the diversity of species and biomass of microorganisms with special microscopes at 250 magnification and can measure length and width of organisms. There is a more precise method of identification through DNA, but it is extremely time consuming and costly.

There are limitations with certain testing methods, such as Petri dish cultures. We cannot have all of these organisms grow under laboratory conditions for testing. Laboratories place organisms in Petri dishes under certain environmental conditions, with specific foods for growing certain organisms, then cover the dishes letting them incubate to see what grows. The environmental conditions are very limited and foods used will grow only certain organisms under those limited conditions. Covering the Petri dishes provides a non-oxygen environment that is a great way to grow pathogens, but very limited for growing a wide diversity of beneficial organisms species or indicating biomass present. These are called plate counts and are useful for growing pathogens, but not a good indication of diversity of species and biomass.

Test Results Indicating Problems

Biology needs to be monitored in the root zones. Nutrients need to be released over time for plants to exist in a healthy growing environment, such as 2 mg of nitrogen per ml. If microbiological tests determine a loss of fungi in your growing system, beware that your plants may be at risk for a fungal disease or other pathogen outbreak. If reports come back that there are too many ciliates (a type of protozoa) in your system, it is an indication that your soil or hydroponics reservoir is starting to go anaerobic. It may be further evidence that a biofilm may be growing somewhere to cause these conditions. Biofilm can also be cleaned up by adding predators (protozoa and nematodes) or adding enzymes. A lack of flagellates and amoebae (other types of protozoa) is also an indicator that anaerobic conditions are present or developing. In hydroponic systems, dissolved oxygen levels should be monitored to determine if aeration needs to be increased to maintain aerobic conditions. If dissolved oxygen levels fall below 6 ppm and anaerobic conditions develop, immediately add more aeration to increase levels above 6 ppm to return to an aerobic environment. If you increase aeration to get your dissolved oxygen levels up, it should clear up anaerobic conditions in hydroponic systems.

Plant Tissue Testing

Plant tissue testing measures the amount of nutrients a plant actually received during the growing season. This is interesting information at the end of the plant growth cycle, but we need testing during the growth cycle. Work is being done that will help predict the final nutrient concentration in plants.

The closer we get to replicating a naturally occurring system, the easier it is going to be to control our plant growing environments.

Combined Foodweb Results

Nature Technologies	(707) 939-7299	Submission Number:	Sample Received	Invoice Number:
Carole Ann Rollins	info@nature-technologies.com	01-016846	07/11/2005	8353
1651 Point Drive				
Sonoma, CA 95476 USA				

Customer Reference	ID	Dry Weight	Active Bacteria	Total Bacteria	Active Fungi	Total Fungi	Hyphal Diameter	TF/TB	AF/TF	AB/TB	AF/AB
Compost Tea											
J1 year	01-100955	N/A	166	176001	69.3	1549	3.25	**0.009**	**0.04**	**0.0009**	**0.42**
J2 Year	01-100956	N/A	70.4	107520	4.77	494	4	**0.005**	**0.01**	**0.0007**	**0.07**
J3 year	01-100957	N/A	70.4	160000	24.2	1163	4.5	**0.007**	**0.02**	**0.0004**	**0.34**

728 SW Wake Robin Avenue Corvallis, OR 97333 USA | (541) 752-5066 | info@soilfoodweb.com | www.soilfoodweb.com 01-016846: Page 1 of 1

TESTS OF A STABILIZED DRY FORM OF COMPOST TEA AFTER ONE YEAR SHELF LIFE: *A year long testing program by Soil Foodweb microbiological testing laboratories of a dry form of a compost tea made by Nature Technologies International, LLC. shows excellent survival and activity. Through a patent-pending process, the tea had been put into specialized containers for one year and then opened and tested for microorganism activity. This is typical of a whole new frontier of compost tea research in the final stages of testing before being released in the commercial market. (Report courtesy of Soil Foodweb, Inc. and Nature Technologies International, LLC.)*

Research on Microorganisms and Interactions

Research over three decades has focused on identifying how many, which type and which set of microorganisms need to be present to protect plant growing environments from diseases, pests and environmental problems. The compilation of data acquired during these 30 years, from over 100,000 samples taken from around the world, have indicated that greater diversity of beneficial microorganisms leads proportionately to the greater likelihood that disease-causing organisms will be out-competed on leaves, stems, roots or soilless media.

When diversity of microorganisms are present, plant-available nutrients will be retained and cycled at rates beneficial to plant needs instead of being leached and lost. Species diversity can reach as high as 25,000 in one teaspoon of compost, worm castings or compost tea. Researchers continue to explore the type and quantity of microbes under what conditions that may be beneficial in fighting harmful organisms, pests and diseases. In addition, there have been studies on types of antibiotic or toxic-compounds produced by microorganisms as another possible mechanism of pest or disease control. A consensus of much of the scientific research indicates the closer we get to replicating a naturally occurring system, the easier it is going to be to control our plant growing environments.

FUTURE SCIENTISTS LEARNING ABOUT COMPOST TEA: *At Live Oak Waldorf School in Meadow Vista, California, children learn about compost tea by using a microscope to see the organisms and then spraying their school yard with the tea in their spray bottles. (Photo courtesy of Sonoma Mountain Institute).*

END NOTES:

1) These inorganic nutrients are different from "organic" based nutrients because of how they are processed. It is often the method of refining ingredients that determines if those ingredients can or cannot be approved for organic growing. No autoclaving (steam, pressure and temperature), or acid treatments are allowed in organic products. There are USDA-NOP agencies that certify or approve the processing and materials used in organic systems. But you do not have to be certified to be growing organically. Just because a label says it is "organic" does not mean it has gone through the approval process of a certifying agency. If you have been certified as an "organic grower" be sure to look for approval of products by a certifying agency on the label. (OMRI, GOCA, etc.)

2) The importance of carbon is that it makes life as we know it possible. Most of the compounds known to science are carbon compounds, often called organic compounds. "Plants absorb the carbon dioxide that makes up around 0.03% of our atmosphere and use the energy of sunlight to build sugar molecules from water and CO_2, a process known as photosynthesis." Living things tend to ingest materials that contain carbon. "There are nearly ten million known carbon compounds and an entire branch of chemistry, known as organic chemistry, is devoted to their study. Many carbon compounds are essential for life as we know it." *(Jefferson Lab, Science Education, Southeastern Universities Research Association (SURA) for the U.S. Department of Energy (DOE). Retrieved December 12, 2005 from: http://education.jlab.org/itselemental/ele006.html.; Fergus Ray Murray, Encyclopedia Britannica, Eric Weisstein's World of Chemistry, Retrieved on December 12, 2005 from: http://fergusmurray.members.beeb.net/carbon.htm.)*

3) "Elements are composed of atoms with characteristic masses. Compounds are composed of atoms combined in small, whole number ratios." An ion is an "atom (or molecule) with missing or extra electrons. Positive ions are cations; negative ions are anions." (*Senese, Fred, Atoms and Ions, General Chemistry Online, Retrieved on December 12, 2005 from: http://antoine.frostburg.edu/chem/senese/101/atoms/index.shtml.*) "An ion is an atom or group of atoms with a net electric charge, having lost (cation) or gained (anion) an electron." (*Retrieved on December 12, 2005 from: http://en.wikipedia.org/wiki/Ion_%28disambiguation%29*).

4) Any ion (salt) has a negative or positive charge that gets pulled apart by water (disassociates in water). The higher the temperature of the water, the faster the salts dissolve. Plants have to have ionic nitrate or ammonium to get nitrogen, but this can be done in an inorganic or organic system very differently. (*Dr. Elaine Ingham, Lecture, Compost Tea in Hydroponics, December 3, 2005, Corvallis Oregon.*)

"Salts contain metal ions. Positive ions are all metallic ions. The endings ".....ite" and ".....ate" mean that they all contain Oxygen (Phosphate, Sulphite, Sulphate, Nitrite, Nitrate, and Carbonate). Where there are two possibilities the ending "....ate" means that there is more Oxygen and the "....ite" means there is less Oxygen." (*Gondar Design Chemistry Index, Positive and Negative Ions. Retrieved on December 12, 2005 from: http://www.purchon.com/chemistry/ions.htm*)

5) *Chemistry Definition of Chelation* – "To combine (a metal ion) with a chemical compound to form a ring." (*The American Heritage Dictionary of the English Language, Fourth Edition, 2003, Houghton Mifflin Company. Retrieved on December 12, 2005 from: http://www.thefreedictionary.com/chelated.*)

"The industrial manufacturing process of Chelation is very inefficient, expensive and painstakingly laborious. Generally only half of the minerals by weight that are put through the process ever really get chelated properly. The other half of minerals remain unchelated in their raw state, but remain mixed together with that fraction which has successfully achieved Chelation, their mixture being sold as Chelated minerals." (*http://www.vaxa.com/library/miner2.cfm*)

6) For a balanced healthy growing system 25,000 active bacterial species are required per gram (teaspoon) of media, 500 micrograms of bacterial biomass per gram (teaspoon) of soilless media (soil, soilless or water), 8000 fungal species, 500 micrograms of total fungal biomass [mg per gram (teaspoon) of soilless media (soil, soilless or water)], and 55% colonization of mycorrhizal fungi. (*Dr. Elaine Ingham, Lecture, Compost Tea in Hydroponics, December 3, 2005, Corvallis Oregon.*)

7) 8,000 Nitrogen molecules are mineralized each day per protozoan. In a good situation, 50,000 individual protozoa are active and functioning. 400,000 Nitrogen molecules are, thus, made available each day in a 24 hour period. This translates into 7 mg of Nitrogen per gram of medium/soil per day is available for your plants to use. But when growing plants they only need 0.2 mg of Nitrogen per gram of medium each day. Other organisms (bacteria and fungi) take up the excess nitrogen that is available and cycle it back into the system. (*Dr. Elaine Ingham, Lecture, Compost Tea in Hydroponics, December 3, 2005, Corvallis Oregon.*)

RESOURCE LIST

Aquatext -- The Free On-line Aquaculture Dictionary provides a table relating Dissolved Oxygen Table (mg/l) to temperature and salinity. http://www.aquatext.com/tables/oxygen.htm.

Acres USA -- A monthly magazine that is the oldest, largest magazine covering eco-agriculture. Acres U.S.A., P.O. Box 91299, Austin, TX 78709 (U.S. mail); or Acres U.S.A., 5321 Industrial Oaks Blvd., Ste. 128, Austin, TX 78735 (UPS, etc.,) http://www.acresusa.com, info@acresusa.com, (800)355-5313, (512) 892-4400, fax (512) 892-4448.

ATTRA (Appropriate Technology Transfer for Rural Areas) -- Sustainable agriculture assistance to farmers, extension agents, market gardeners, agricultural researchers, and other agriculture professionals in all 50 states. The latest in sustainable agriculture and organic farming news, events and funding opportunities. National Sustainable Agriculture Information Service, P.O. Box 3657, Fayetteville, AR 72702, 800-346-9140, 800-411-3222, http://www.attra.org.

Bio-Integral Resource Center, BIRC, PO Box 7414, Berkeley, CA 94707, Phone: (510) 524-2567, Fax: (510) 524-1758, http://www.birc.org/, Email: birc@igc.org, We specialize in finding non-toxic and least-toxic, integrated pest management (IPM) solutions to urban and agricultural pest problems. Publications on IPM and alternative solutions to pest problems. Magazines include "Common Sense Pest Quarterly" and "The IPM Practicioner".

Convention on Biological Diversity -- Dedicated to promoting sustainable development. The Convention recognizes that biological diversity is about more than plants, animals and micro organisms and their ecosystems – it is about people and our need for food security, medicines, fresh air and water, shelter, and a clean and healthy environment in which to live. Phone: +1 514 288 2220, http://www.biodiv.org/default.shtml, SCBD Secretariat, Email: secretariat@biodiv.org.

Dissolved Oxygen Calculator -- Quick DO calculator where you can plug in both water temperature and barometric pressure (a more precise variable than altitude alone, because it varies with both altitude and weather conditions). Fred Senese senese@antoine.frostburg.edu, http://antoine.frostburg.edu/chem/senese/101/solutions/faq/predicting-DO.shtml.

Earth Fortification Supplies -- The products offered support the Compost and Compost Tea industry, with an emphasis on biological and sustainable plant and soil health. Products include: Bacterial foods, fungal foods, compost, instruments, literature, microscopes, and tea brewers. Address: 1750 S.W. 3rd Street, Suite K., Corvallis, OR 97333. 541-257-2612. Email: info@earthfortification.com. Website: www.earthfortification.com.

The Ecological Farming Association (EFA) -- EFA supports a vision for our food system where strengthening soils, protecting air and water, encouraging diverse ecosystems and economies, and honoring rural life are all part of producing healthful food. Our innovative programs bring together growers, consumers, educators, activists, and industry related businesses to exchange the latest advances in sustainable food production and marketing. Address: 406 Main Street Ste. 313, Watsonville, CA 95076, ph. 831-763-2111, fax. 831-763-2112, http://www.eco-farm.org, info@eco-farm.org.

ETC group -- Dedicated to the conservation and sustainable advancement of cultural and ecological diversity and human rights. ETC group supports socially responsible developments of technologies useful to the poor and marginalized and it addresses international governance issues and corporate power. Action group on Erosion, Technology and Concentration. ETC Group Headquarters, 431 Gilmour St, Second Floor, Ottawa, ON Canada K2P 0R5, Tel: 1(613)241-2267, Fax: 1(613)241-2506 http://www.etcgroup.org.

Hendrikus Schraven – He has been applying, innovating and researching with compost tea since the first brewers were available. Services and Program: Fertilizer program, compost tea program, soil installation, all organic landscaping, and seminars. Products: fertilizer blends, compost tea, specialty fertilizers, liquids, soil amendments, pest and weed defense, and essential soil. Websites: www.hendrikusorganics.com (landscape construction company) and www.soildynamics.com (soils & erosion control company).

The International Compost Tea Council -- A group of industry leaders in compost tea doing research and development, documenting the uses of compost tea. The ICTC provides open, informed scientific and research data in user-friendly format; educational information, and a forum through which the compost tea industry can be heard as a unified voice in protecting and preserving this valuable tool for all users. Address: ICTC, 14150 NE 20th St, Suite 293 Bellevue, WA 98007, Website: http://www.intlctc.org.

Managing Wholes -- This grassroots site is about management for simultaneous social, financial, and ecological benefit. Our goal is to promote change toward a future that works. Firsthand reporting about what people are learning from managing wholes rather than parts in agriculture, ranching, forestry, and community development. Peter Donovan, 63237 Pine Tree Rd., Enterprise, Oregon 97828, Wilma Keppel, 1058 Pomona Ave., Albany, California 94706, 510-526-1234, 541-432-0710, http://managingwholes.com.

Mycorrhizal Applications Inc. -- Manufacturer-distributor-retailer of VA and Ecto mycorrhiza inoculums. Pathogen free with constant and rigid quality control measures. (541) 476-3985, (866) 476-7800. Website: www.mycorrhizae.com

Organic Consumers Association -- The Organic Consumers Association (OCA) is an online and grassroots non-profit public interest organization campaigning for health, justice, and sustainability. The OCA deals with crucial issues of food safety, industrial agriculture, genetic engineering, children's health, corporate accountability, Fair Trade, environmental sustainability and other key topics. Address: 6771 South Silver Hill Drive, Finland, MN 55603, 218-226-4164, Fax: 218-353-7652, http://www.organicconsumers.org.

Organic Farming Research Foundation (OFRF) -- The Organic Farming Research Foundation is a non-profit whose mission is to sponsor research related to organic farming practices, to disseminate research results to organic farmers and to growers interested in adopting organic production systems, and to educate the public and decision-makers about organic farming issues. Address: P.O. Box 440, Santa Cruz, CA 95061, phone: 831-426-6606, fax: 831-426-6670, http://www.ofrf.org/.

Oregon Tilth, Inc. -- A nonprofit research and education membership organization dedicated to biologically sound and socially equitable agriculture, certifying organic farmers, processors, retailers and handlers throughout Oregon, the United States, and internationally. Address: 470 Lancaster Dr., NE, Salem, Oregon 97301, (503) 378-0690, FAX (503) 378-0809, http://www.tilth.org, organic@tilth.org.

Oregon Sustainable Agriculture Land Trust (OSALT), OSALT is a charitable nonprofit organization and has an integrated, four-fold purpose: to protect rural and urban agricultural lands and keep them in agricultural production; to make those lands available for the use of future generations of growers; to conduct research on those lands into the sustainable production and distribution of agricultural bounty; and to provide education focused on living sustainably. Address: PO Box 1106, Canby, Oregon 97013-1106, http://www.osalt.org info@osalt.org.

Natural Resource Ecology Laboratory (NREL) -- The driving goal of NREL science is to increase the fundamental understanding of our environment, through collaborative application of ecology and ecosystem science principles which contribute to policy and management decisions promoting sustainability. NREL is an interdisciplinary ecosystem science research center within the College of Natural Resources of Colorado State University. Address: NREL Dept., 1499, 200 West Lake, Ft. Collins, CO 80523-1499, Tel: (970) 491 5571 - Fax: (970) 491-1965, http://www.nrel.colostate.edu.

Nature Technologies International, LLC. – Manufacturer of stabilized Compost Tea with shelf life of active organisms for over one year due to new patented technology. Manufactures and distributes Stainless Steel Compost Tea Brewers, Compost Tea Ingredients, Worm Castings, Humic Acid, Mycorrhizae and Sea Kelp, Brand Name "Nature's Solution". Licenses Compost Tea Brewing Service Centers and provides consulting services and seminars. Address: P.O. Box 1519, Novato, CA 94947, 415-898-5895, Email: naturetech@earthlink.net. Website: www.nature-technologies.com.

Rincon-Vitova Insectaries, Inc. -- Publishers of "Catalog of Beneficials" based on ecologically-based agriculture where beneficial organisms are encouraged or released to control pests. 800-248-2847, bugnet@rinconvitova.com. Address: Box 1555, Ventura, CA 93002-1555 and 3891 Ventura Ave (rear), Ventura, CA 93001-1243. 805 643-5407 800-248-2847 Fax 805 643-6267. Email bugnet@rinconvitova.com. Website: http://www.rinconvitova.com

Soil Foodweb Laboratories – At soil foodweb we measure the quality and quantity of soil microorganisms and guide you to build the life in your soil and grow healthier plants. Laboratory location are around the world. Look at the website for a location nearest you. Address: 71750 S.W. 3rd Street, Suite K, Corvallis, OR 97333. 541-752-5066. Email: info@earthfortification.com. Website: www.earthfortification.com.

The Sonoma Mountain Institute -- The Institute is a non-profit organization based in Sonoma County, California dedicated to caring for natural systems -- restoring and revitalizing natural landscapes, savannah Grasslands, Oak Woodlands, to offset compaction and degraded soil from overgrazing in the past. Phone 415-302-4332.

Sustainable Studies Institute -- The Sustainable Studies Institute is a non-profit organization focused on research and education. Soil microbiology, soil ecology, nutrient cycling, and biological control of soil borne pathogens are some of the focus areas of study. The organization offers workshops and seminars on soil biology, compost, compost tea, and light microscope methods developed by Dr. Elaine Ingham based on the Soil Foodweb principles. If you would like to contact them to book a speaker, or attend a seminar, please contact the Sustainable Studies Institute at: Address: 1750 S.W. 3rd Stret, Suite K, Corvallis, OR 97333. 541-752-2614. Fax 541-752-5142. Email: info@sustainablestudies.org. Website: www.sustainablestudies.org.

ABOUT THE AUTHORS

Dr. Elaine Ingham, Ph.D, is President of Soil Foodweb Inc. an international laboratory system, that assesses the balance of bacteria, fungi, protozoa, nematodes and mycorrhizal fungi in all materials. The major emphasis of her work is to return health to soil, so that natural nutrient cycling and disease suppression mechanisms are present, allowing the desired plants to grow without requiring use of pesticides or inorganic fertilizers. Sustainable Studies Institute, 728 S.W. Wake Robin, Corvallis, OR 97333-1612, (541) 257-2614. info@ sustainablestudies. org, www. sustainablestudies.org

Dr. Carole Ann Rollins, Ph.D., is co-owner of Nature Technologies International LLC with her husband James Eddington, manufacturing Compost Tea Machines, Ingredients, and a Compost Tea with a one year shelf life with active organisms. The company offers a full line of organic products under the trademark Nature's Solution (Compost Tea, Sea Kelp, Ancient Humate, Mycorrhizae, and Worm Castings). Carole offers private consultations and seminars. Nature Technologies Internationnal, 14 Tilden Drive, Novato, CA 94947, 415-898-5895, naturetech@earthlink.net, www.nature-technologies.com.